THE MATTER OF LIFE

Student Manual

DEVELOPED BY
EDUCATION DEVELOPMENT CENTER, INC.

KENDALL/HUNT PUBLISHING COMPANY
4050 Westmark Drive P.O. Box 1840 Dubuque, Iowa 52004-1840

This book was prepared with the support of National Science Foundation (NSF) Grant ESI-9255722. However, any opinions, findings, conclusions and/or recommendations herein are those of the author and do not necessarily reflect the view of NSF.

Library of Congress Catalog Card Number: 96-80037

ISBN 0-7872-2206-2

Printed in the United States of America
10 9 8 7 6 5 4 3 2 1

EDC Education Development Center, Inc.

CENTER FOR SCIENCE EDUCATION

Dear Students:

Welcome to *Insights in Biology*. In this module *The Matter of Life,* you will be exploring the characteristics that define a substance as having "life." What processes do living things carry on that distinguish them from nonliving matter? How do organisms carry on these processes? One of the major themes of this module is that there is unity among the diverse forms of life. The characteristics of life and the processes that enable life to continue are similar among all forms of life. This is a simple yet profound concept.

Glance through the pages of this book. Your first instinct is correct: This is not a traditional biology textbook. Although textbooks provide a good deal of useful information, they are not the only way to discover science. In this Student Manual, you will find that chapters have been replaced by Learning Experiences that include readings and activities. The readings consist of materials from magazines, newspapers, books, and original writings. The activities include laboratory experimentation, role playing, concept mapping, model building, simulation exercises, and research projects. This inquiry-based, modular curriculum emphasizes both the processes of science and the importance of deep understanding of biological concepts and the connections among these concepts.

One of our main goals is to engage you in the excitement of biology. The study of biology is much more than facts. It is a discipline that is as alive as the subjects it portrays: new questions arise, new theories are proposed, and new understanding are achieved. As a result of these new insights, technologies are developed which will impact your everyday lives and the kinds of decisions you need to make. We hope that this curriculum encourages you to ask questions, to develop greater problem-solving and thinking skills, and to recognize the importance of science in your life.

Insights in Biology Staff

55 CHAPEL STREET
NEWTON, MASSACHUSETTS 02158-1060
TELEPHONE: 617-969-7100
FAX: 617-630-8439

TABLE OF CONTENTS

LEARNING EXPERIENCES

1 **LIVING PROOF** 1

Living or Nonliving?—Activity
Aliens—Reading

2 **WHICH WAY DO THEY GO?** 7

Where Is the Light?—Activity
What's Your Response?—Reading

3 **EVERYTHING UNDER THE SUN** 15

Making the Connections—Activity
You Light Up My Life—Activity
How Does Your Garden Grow?—Reading
Sunlight Becomes You—Reading

4 **FEEDING FRENZY** 27

What Am I Eating, Anyway?—Activity
The Missing Ingredients—Reading

5 **THE LEGO OF LIFE** 39

Madam, I'm Atom—Reading
A Mere Six Ingredients—Activity

6 **TURNING CORN INTO MILK: ALCHEMY OR BIOCHEMISTRY?** 47

Corn and Milk: So Different Yet So Similar—Activity
Haven't I Seen That Carbon Somewhere Before?—Reading

7 **A Breath of Fresh Air** 57

What Goes In...—Activity
...Must Come Out—Activity
Energy for All—Reading

8 **Reap the Spoils** 71

Home Sweet Milk—Activity
Techniques—Reading
Continuing to Reap the Spoils—Activity

9 **How Did It All Begin?** 79

The Bubbling Cauldron—Activity
The Spark of Life—Reading

10 **Night of the Living Cell** 87

Our Bodies, Our Cells—Reading
The Whole Cell and Nothing But the Cell—Activity
As the Cell Turns—Reading
An Organelle With A View—Reading

11 **The Great Divide** 99

Soaking It All In—Activity
Divide and Conquer—Reading

12 **Return to Reap the Spoils** 109

Reaping the Spoils—Activity

13 **Where Will It All End?** 113

When Organisms Die—Reading
"What If It Works?"—Reading

14 **Suspended Between Life and Death** 119

The Problem Defined—Activity
A Difficult Choice—Reading
Is Life Extended?—Activity

Appendix

A **Nutrient Content of Food** 127

B **Glossary of Terms** 135

Living Proof

PROLOGUE Imagine yourself sitting beside a pond in early autumn. A maple tree shades you from the hot sun, while blades of grass and newly fallen leaves form a soft cushion under you. A fish jumps, breaking through a green film that has recently formed on the pond's surface. Birds sing, and bees and butterflies dart in and around the colorful flowers, collecting the last nectar of summer. You toss a rock in the water, then watch concentric rings flow outward from the spot where the rock disappeared.

Which of the things described above are living? This seemingly trivial question is not necessarily easy to answer. What are the characteristics that we can use to define something as living? Biologists have so far identified over three million species of living things, occupying every nook and cranny of Earth's surface (not to mention a vast array of species that once occupied the planet but are now extinct). Their diversity in size, structure, and ways of living is astonishing, yet all share specific and definable characteristics that distinguish them from nonliving things.

Scientists are often confronted with the challenge of determining whether something is living or nonliving. Samples from geological expeditions, from the ocean bottom, from other planets are brought back and analyzed for characteristics of life as we know it. Searching for signs of life can be quite a difficult task. Look at Figure 1.1. Is the plant living? How do you know?

Figure 1.1
Is the plant living? How do you know?

Living or Nonliving?

INTRODUCTION You will observe a sample of an unidentified substance and, using any approach you can devise, determine whether or not the substance is living. It is extremely important that you base your decision only on what you can observe in your sample and that you can explain your decision.

MATERIALS NEEDED

For each pair of students:

- 2 hand lenses
- 1 glass stirring rod or dissecting needle
- 1 sample of unidentified substance

For the class:

- microscopes

PROCEDURE

***SAFETY NOTE:** When you finish examining the materials, wash your hands thoroughly using laundry soap or other strong soap.*

1. Observe the sample closely, using available materials. Use your own observations to determine whether or not the sample is living. Record in your notebook your reasons as to why you believe it to be living or nonliving.
2. Use the class list of the characteristics of life and describe how you used those characteristics to help determine whether or not you think your sample is living.
3. **STOP & THINK** What questions would you need to ask in order to gain more knowledge about whether your sample is living or nonliving?
4. **STOP & THINK** How might you go about answering your questions?

READING

Aliens

THE SEARCH FOR EXTRATERRESTRIAL LIFE

Humans long ago learned that the planets in our solar system are physical heavenly bodies made of matter, perhaps similar in some ways to our Earth. Knowledge that stars are also similar to our home star, the Sun, has often provoked people to wonder "Is there life elsewhere or are we alone in the universe?" The question remains, but methods for attempting to answer it have become quite sophisticated. One approach which has been carried out has been to test actual samples from other planets. In 1975, two spacecrafts—Viking 1 and Viking 2—were sent to Mars with the mission of determining whether life, as we know it, was present on the red planet. To ensure that no contamination of Earthly life-forms occurred, the spaceships were sterilized; this was done by heating the Vikings to 113°C and then sealing them inside a shield to keep out potential hitchhiking microorganisms.

Once the spacecraft landed on Mars, samples were collected and subjected to a variety of different analytical tests used to identify the presence of living organisms. These tests detect whether the samples:

- have substances which contain the element carbon;
- produce food of some kind;
- show evidence that food is consumed;
- demonstrate changes in the gas content.

Figure 1.2
The Viking spacecraft landing on Mars. (Not drawn to scale.)

When the Martian samples were analyzed, no evidence of carbon-containing substances was found nor was food production or consumption observed. However, the gas content did show significant change. Whether this was caused by the presence of living things or by some strange chemical reactions on the planet was not clear.

ANALYSIS

Write responses to the following in your notebook:

1. Describe what the tests for detecting life forms tell you about what it means to be alive.
2. What other indications or characteristics of life were scientists looking for?
3. Why do you think air and food are important to staying alive?
4. What is meant when scientists speak of not finding life "as we know it"?
5. Have you ever brought a wild animal home from the woods, picked up a snail at the beach, or found a wounded bird and brought it home? Were you able to keep it alive? What did you do to try to keep it alive?
6. What is the difference between living, alive, nonliving, and not alive?

EXTENDING IDEAS

- Set up a habitat for your organism. Research the natural habitats of the organism and any requirements it would need to stay alive. Determine what materials you would need, and what resources you would need for the organism to survive.
- Read or write your own story, poem, or song, examine a piece of art, or create a collage which develops the theme of what life is. In a short essay, explain what question this piece asks about life and how the question is answered for the reader, listener, or observer.
- Describe an experience you have had in which you found something and tried to determine whether it was living or nonliving.

ON THE JOB

SCIENCE WRITER Do those science-related articles on television, radio, magazines, or newspapers capture your interest? Science writers combine their science background and ability to express ideas clearly to translate technical scientific research and language into language that is more easily understandable for the general public. In researching a story idea or developing a story, the science writer might read technical research papers and interview researchers (experts in the field). Many hours of background research and preparation go into the 3–4 minute segment you see or hear on television and radio or read in a column in a newspaper or magazine. Most science writers have a minimum of a college degree in a scientific discipline (although some have a liberal arts degree), and sometimes a graduate level degree either in science, science journalism, or communications. Beginning in high school, classes such as life science, chemistry, physics, English composition, or communications are useful. The American Association for the Advancement of Science (AAAS) offers a Mass Media Science Fellowship.

LABORATORY ASSISTANT Are you organized and detail oriented? Laboratory Assistants are responsible for lab tasks and some parts of experiments. On any given day they might be making detailed observations, analyzing data, interpreting results, or writing summaries of the protocols for an experiment. Other tasks might include maintaining lab equipment, calibrating instruments, and monitoring the inventory level for lab supplies. Laboratory assistants might work in laboratory research at a private company, government research facility, or for a field research project. A minimum of an associate's degree is required with 0–2 years of laboratory experience. For those with more advanced degrees, there are additional laboratory positions available. Beginning in high school, classes such as computers, English, biology, mathematics, chemistry, or physics are useful.

Which Way Do They Go?

PROLOGUE What do organisms need to stay alive? What are some basic resources? How might organisms obtain any of these resources? Being dependent on their environment for the resources they require to maintain life, organisms need ways to recognize and respond to these resources. For example, organisms can detect the presence of required nutrients and respond by moving toward them. You probably respond in this way when the aroma of a large pizza enters your environment.

As long as conditions remain constant, organisms may not need to respond any differently than usual. What happens to an organism when its environment changes in a significant way? If a change occurs within an organism's environment, making the surroundings less hospitable, the organism will seek more favorable conditions. If the change is catastrophic and an organism can neither reach more favorable conditions nor adjust to the changed environment, it will die.

Where Is the Light?

INTRODUCTION In the following activity, you will investigate the response of single-celled, photosynthetic organisms, euglena *(Euglena)*, to a change in their environment.

MATERIALS NEEDED

For each pair of students:

- 2 pairs of safety goggles
- 1 euglena phototaxis kit or culture
- 1 tall, straight, clear jar with screw cap

- spring water or filtered pond water
- 1 sheet black construction paper
- 1 eyedropper
- 1 microscope slide and coverslip
- 1 rubber band
- 1 scalpel or utility knife
- 1 sheet of cardboard (or other thick paper) as cutting surface
- 1 paper towel
- 1 dissecting needle
- access to a compound microscope

▶ PROCEDURE

SAFETY NOTE: *Always wear safety goggles when conducting experiments.*

1. Pour about 15 mL (1 oz) of euglena culture into a straight-sided glass jar. Add spring water or filtered pond water to fill the jar to the top and then screw on the cap lightly; do not tighten. Observe and record the appearance of the tiny euglena in your notebook.

SAFETY NOTE: *Always use caution with sharp objects.*

2. Use a scalpel to cut a piece of black construction paper of the same height as the jar and wide enough to wrap around the jar once, with the ends slightly overlapping (see Figure 2.1).
3. Decide on a geometric shape, a letter of the alphabet, or other interesting design that is about 2 cm (1 in) in diameter. Cut the design out of the center of the paper. (Place construction paper on a flat cutting surface and, using the scalpel, cut out the design.)
4. Wrap construction paper around the jar and secure it with a rubber band.

Jar
Rubber band
Hi
Black construction paper

Figure 2.1
Setup for euglena investigation

CAUTION: *Do not allow the jar to become overheated.*

5. Place the jar in a well-lit location so that the cut-out design is facing the light.
6. STOP & THINK What are you hoping to find out about euglena by carrying out this experiment? (That is, what question is being asked in this investigation?)
7. STOP & THINK Scientists often create a hypothesis to help plan an experiment. A hypothesis is an opinion or guess as to the probable answer to a question. In formulating a

hypothesis, scientists rely on prior knowledge, observations, and experience. Some call this an "educated guess." Once a hypothesis is made, an experiment can be designed to determine whether the hypothesis is reasonable and correct. For example, an investigator may pose the question "what resources do plants require in order to sustain life?" A hypothesis would then state "Plants require sunlight, water, and air in order to sustain life." The investigator then designs an experiment or experiments to test this hypothesis. The experimental design is generally framed as "If I do *this* (remove light, for example) then *that* (expected result) should occur." Based on your observations of euglena, your understanding of resources, and any prior knowledge you have, create a hypothesis about the investigation you just set up.

8. Obtain a second, smaller sample of euglena for microscopic examination. Prepare a wet mount slide as follows:
   - Using an eyedropper, place a drop of the culture in the middle of a microscope slide.
   - Place an edge of a coverslip on the slide near the specimen. Carefully and slowly lower the coverslip over the specimen using a dissecting needle or other support. (Do not drop on top.) (See Figure 2.2.)
   - Remove any excess fluid by touching the edge of a paper towel to the outer edge of the liquid under the coverslip. The coverslip should lie flat on the microscope slide; Press gently on the coverslip to remove any air bubbles.

Figure 2.2
Preparation of microscope wet mount slide

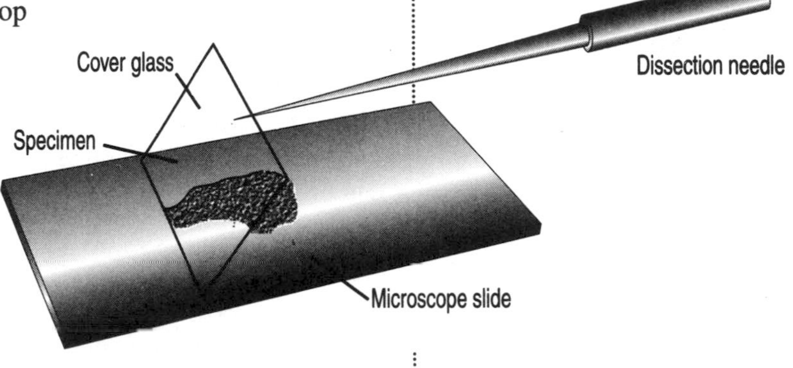

9. Place the slide carefully on the center of the microscope stage and observe the euglena under different powers of magnification.
10. Draw or describe any features of euglena that you observe with the microscope.
11. After making your observations, dispose of your euglena sample in the site designated by your teacher.
12. After 24 hours, gently pick up your jar setup and carefully slide the paper down. Be sure not to disturb the jar's contents. Describe the results of your experiment and respond to the following Analysis questions in your notebook.

ANALYSIS

1. Do the results of the experiment support or contradict the prediction from your hypothesis? Explain your answer.
2. What response have the euglena exhibited? Why do you think they have responded this way?
3. What features of euglena enable them to respond in this way?
4. Do you think other organisms would respond the same way? Why or why not?
5. What do you think would happen if you placed the jar in complete darkness overnight? For two weeks?

WHAT'S YOUR RESPONSE?

All living things have the capacity to react to their environment. While some of these reactions may be so subtle as to be invisible, others may result in enormous changes in the responding organisms.

This ability to respond to the environment in a controlled way is one of the characteristics of life. When interacting with the world, all organisms must be able to sense and respond to continually changing conditions around them.

Even *Escherichia coli*, a single-celled bacterium, can respond to changes in its environment. Inside the cell, several proteins continually monitor the level of nutrients in the surrounding watery environment where the *E. coli* lives. If the nutrient levels remain steady or increase, the bacterium will move about in a random fashion; but when levels begin to fall, the proteins interact with the flagellum (the whiplike part of a bacterium that enables it to move) causing the *E. coli* to move toward higher concentrations of nutrients.

Dictyostelium discoideum, a slime mold, uses a different strategy, connected with its more complex life cycle. As long as nutrients are available and conditions remain constant, the slime mold can exist as a group of independent single-celled organisms called amoebas. When the food supply available to *D. discoideum* begins to dwindle, the cells stop dividing and the individual amoebae start moving toward one another. Eventually they clump together to form a slug-like, multicellular structure that moves along the surface of the soil. The slug develops into a new structure, called a fruiting body, that resembles a ball on a stalk. This fruiting body is the amoeba escape vehicle. Cells in the ball at the top of the fruiting body mature into spores, or reproductive cells, each of which is capable of forming a new amoeba. These spores are scattered across the surface of the soil by wind, animals, or rain, and in some cases into more favorable environmental conditions where nutrients are available (see Figure 2.3). The spores then give rise to individual amoebae which continue to grow and divide independently of one another, thus continuing the cycle and ensuring the survival of this slime mold through its offspring.

Plants have ways of responding to stimuli in their environment as well. Sometimes plants are mistakenly regarded as nonliving, partly because they do not display responses to their environment as speedily or visibly as animals do. Animals seem to respond to their environment quickly and decisively. In comparison, trees and flowers may seem inanimate because they appear motionless; yet, plants can and do respond to their environment. For example, grass responds to the amount of water in the soil by adjusting the depth of its roots: infrequent light rain will cause the roots to spread out close to the surface to capture the small amount of moisture; regular heavy rain, on the other hand, allows grass to sink its roots deeply, as the water penetrates deeper into the soil.

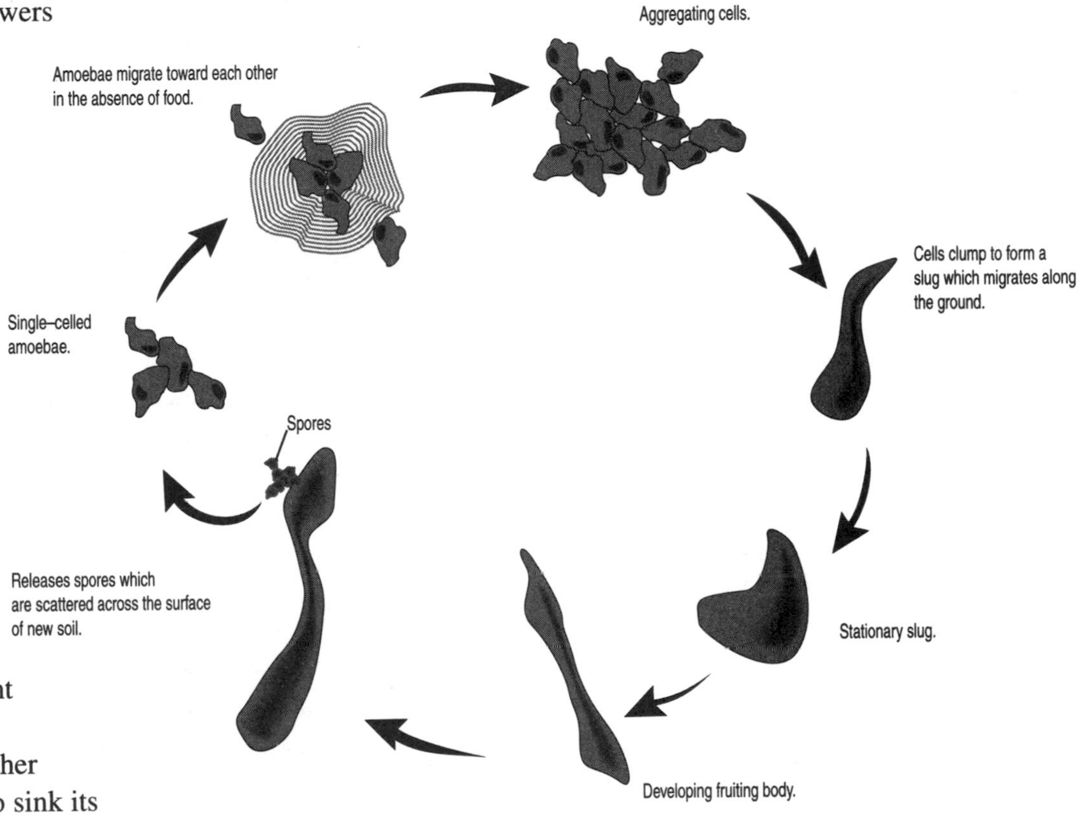

Figure 2.3
A slime mold's responses to dwindling resources in the environment.

Humans, like most animals, are designed to move in response to detailed awareness of their environment. The bulk of the body is dedicated to sense, reaction, and motion. Cells in our eyes react to light, cells in our ears sense a myriad of sounds. These and other sensory organs provide data that are sent as messages to the brain, which processes these messages (within thousandths of a second) and then, in turn, sends out signals to other parts of the body, such as to our muscles and organs, to respond appropriately. For example, our sense of smell will guide us to the source of hunger-satisfying foods.

Organisms respond to their environment for many reasons besides obtaining resources. The responses include three important sets of factors:

- various physical factors allowing the organism to survive and reproduce, including: level of salinity, pH, temperature, humidity, type of soil, and others
- biological factors that may interact with the organism, including: predators, parasites, and any competitors for the same resources
- behavioral factors including the organism's feeding habits and its behavior when interacting with other organisms.

Some plants respond to factors in their environment in observable ways. Have you ever touched a sensitive plant? Heat from your hand causes it to react by folding in its leaves. Tulip petals open wide during the day and then close up each night. Evening primroses on the other hand open at dusk and close again when the hot sun begins to shine on them.

Animals have senses with which they perceive such conditions in their environment as light, sound, smell, temperature, and air, and they respond by actively moving toward hospitable environments and away from hostile environments. The striped bass is one example of an animal that requires a specific temperature as an optimal environmental condition. Fish cannot control the temperature of their bodies; instead, they depend on the temperature of the water around them. Their metabolic processes (the ability to break down and transform food into useful nutrients and energy) function best at a specific temperature. The fish move toward that temperature zone in the water, sensing changes in their environment and moving when necessary toward the appropriate temperature (see Figure 2.4).

Figure 2.4
Adult bass need temperatures below 25°C (77°F) in order to survive. When the sun heats the surface water, the bass move into areas that have more suitable conditions. If these suitable environments become too crowded, the bass may be forced into water that is warmer than they can tolerate, and they die.

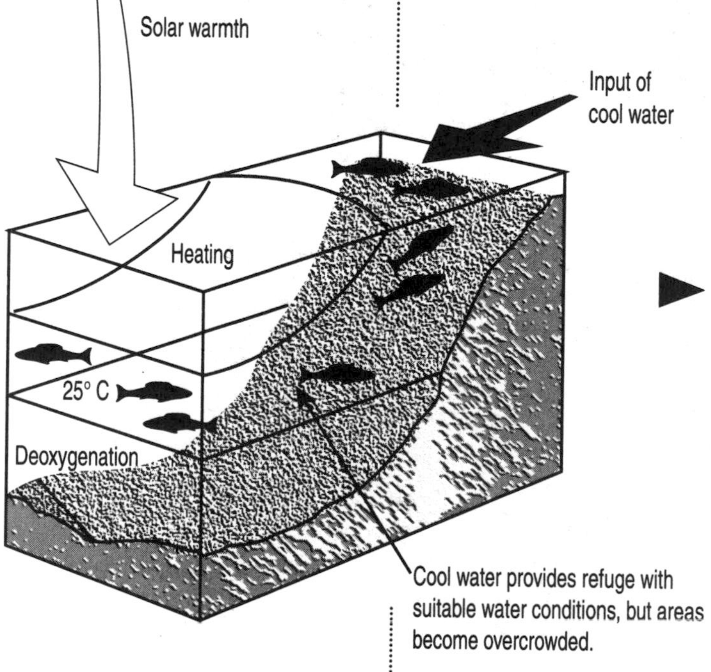

Solar warmth

Input of cool water

Heating

25° C

Deoxygenation

Cool water provides refuge with suitable water conditions, but areas become overcrowded.

The ability to respond is a necessary characteristic for survival. That ability alone is not enough to define life; without it, however, organisms would be unable to get necessary nutrients and resources or to seek out vital environmental conditions.

ANALYSIS

1. List three ways (not found in the above reading) in which organisms adjust in order to find and use resources in their environment.
2. Describe why the responses you listed are important to that organism.
3. What might cause an environment to change? Describe three changes which might alter the availability of resources in an environment.

4. Predict how these changes might affect the survival and behavior of organisms living in that environment. Explain your predictions.
5. How do the other responses described in the reading help an organism to maintain life?
6. Based on the results of the "Where Is the Light?" experiment, describe the relationship between the response of an organism and the requirements for maintaining life.

EXTENDING IDEAS

Organisms communicate within themselves and with each other in a fascinating number of ways: language, sophisticated mating dances, chemical signaling, and so forth. The purpose of these varying types of communication is to allow organisms to communicate, enabling them to survive. Investigate this in more depth by researching one or more of the following:

- the biological basis of social behavior among bees or ants
- the development of language
- the biological basis of pheromones (and their role in human behavior)
- intracellular communication
- hormones
- the nervous system

ON THE JOB

LANDSCAPE ARCHITECT Did you ever think of combining art and science into one career? Landscape architects design and plan outdoor spaces that make best use of the land and respect the needs of the natural environment. A landscape architect needs to be a scientist, an architect, and an engineer rolled into one in order to consider the desires and needs of the client, observe and map the land that is being developed, consider existing features such as trees and shrubs or buildings, identify the shady and sunny areas, and test the composition of the soil. Landscape architects design, plan and make recommendations for large and small communities, parks, playgrounds and plazas, recreation facilities, waterfronts, hotels and resorts, shopping centers, roads, public housing, and nature conservation areas. A college degree or graduate study is required. Classes such as English composition and literature, history, government, sociology, psychology, biology, mathematics, and mechanical and freehand drawing are needed.

AGRICULTURAL SCIENTIST Did you know that many farmers work with scientists to make farming easier and more profitable? While agricultural scientists are concerned with all aspects of living organisms and the relationships of plants and animals to their environment, they focus on plants, insects, or farm animals. Their work can be done in a laboratory, in the field, or a combination of both places. Some agricultural scientists may analyze soil to find a way to increase crop production or decrease soil erosion; others experiment in breeding plants to improve resistance to weather, disease, or insects, or to produce a higher crop yield. Still others apply their research to the food or agricultural industries by working on flood control or soil erosion, or they use their knowledge to make farming easier and more profitable. Positions as laboratory assistants or technicians may be available for those with high school diplomas; as entry level testing or inspecting technicians, sales or service representatives for those with a college degree; those with a master's degree can do basic or applied research; and independent research or college level teaching jobs are available to those with a Ph.D. At every level, classes in subjects such as chemistry, math, physics, and biology are useful.

Everything Under the Sun

PROLOGUE **W**hat kinds of substances in the environment are essential for life? In Learning Experience 2, you determined that euglena respond to light. What other resources in the environment are required to maintain the characteristics of life in organisms?

In this learning experience, you continue to identify resources that are essential to survival, specifically in the environment of a plant, and then you explore how a plant uses these resources.

Making the Connections

INTRODUCTION What resources do plants require to maintain life? Where do you think plants get each of these resources? Construct a concept map which illustrates the resources plants require and how the resources might be obtained and used. To create a concept map, follow these basic rules:

TASK

1. Identify the concepts to be mapped. Create a list.
2. Decide which concept is the main idea. Look for ways to classify the remaining concepts. Rank them from general to specific.
3. Pick linking words for the map that identify the relationships between the concepts. Linking words should not be concepts themselves.
4. Start constructing a map by branching one or two general concepts from the main concept. Add other, more specific concepts to the general ones as the map progresses.

5. [Put each of the concepts in a "box" or "circle." Use lines to connect the concepts. Write a linking word on the line that tells why concepts are connected.]
6. Look for opportunities to draw cross-linkages to connect concepts from different branches of the map.

Excerpted from a workshop entitled Concept Mapping, A Strategy to Help Students Learn How to Learn, *Dr. Laine Gurley Dilger, D.C. Heath and Co.*

ACTIVITY

You Light Up My Life

INTRODUCTION In the following investigation you will use radish plants to investigate the resource requirements of a plant and find out what happens if a plant does not obtain the resources it needs. You then use this information to examine the biochemical processes by which a plant obtains what it needs for survival.

What happens within a plant when it obtains the resources it requires? You may have already studied or examined photosynthesis in other classes. *Photosynthesis* is the process in which plants use the resources of sunlight, carbon dioxide (CO_2), and water (H_2O) to fulfill their needs for energy and food. The components of the word—"photo" meaning light and "synthesis" meaning putting together—refer to its functions.

What structural components of a plant are important in obtaining and using resources? Figure 3.1 is a drawing of a plant. Below the soil surface, a highly

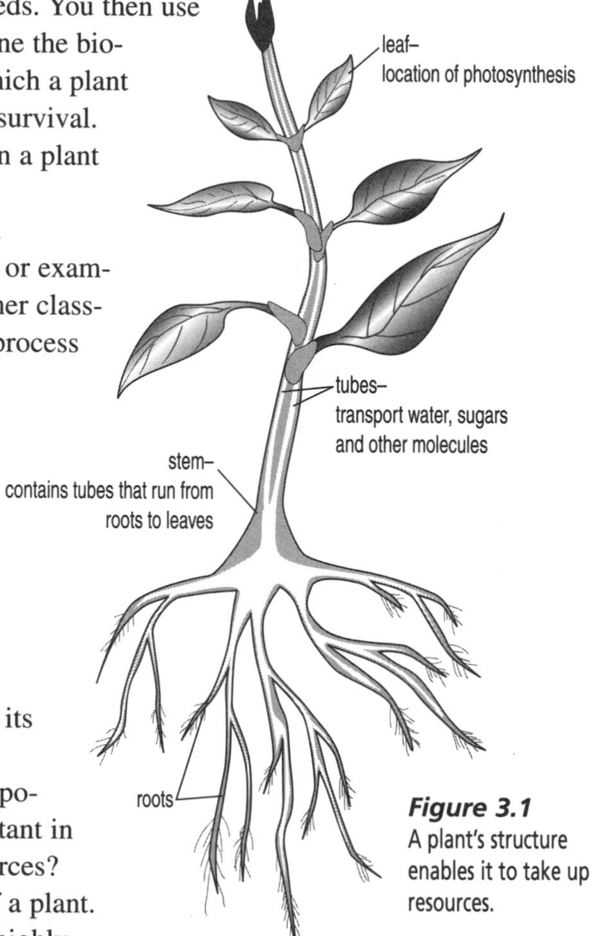

Figure 3.1
A plant's structure enables it to take up resources.

branched root system brings in water and minerals from the soil and anchors the plant in the ground. The stem contains a system of tubes that begins in the root and runs all the way to the top of the plant. One set of tubes (*xylem*) in the stem conducts water and dissolved materials drawn from the soil up to the leaves (Figure 3.2). Another set of tubes (*phloem*) in the stem transports sugars, the products of photosynthesis, and other molecules throughout the plant.

The leaf is the major site of photosynthetic activity. Carbon dioxide enters the leaf through tiny holes in the leaf's surface. Water travels from the roots through the stem and enters the leaf. Special cells in the leaf can absorb energy from the sun using a molecule called *chlorophyll*. The water, carbon dioxide, and energy from the sun (*solar energy*) are all essential components in the photosynthetic reaction and come together in special cells in the leaf. As you work with your radish plants, be sure you are thinking about the relationship between the plant parts and the resources it needs.

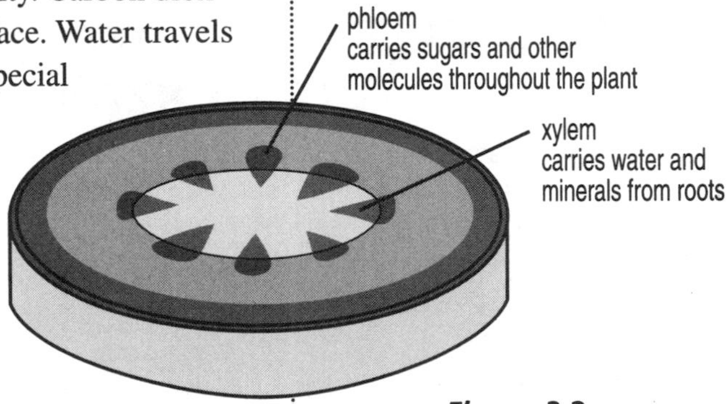

Figure 3.2
The cross section of a stem showing the phloem and xylem.

MATERIALS NEEDED

For each group of four students:

- 4 small pots of radish plants, each grown under one of the following conditions:
  - in the light, under normal conditions
  - in the dark, under otherwise normal conditions
  - in a closed environment, under otherwise normal conditions
  - in a closed environment in the absence of carbon dioxide (CO_2)
- 4 safety goggles
- iodine solution
- 4 petri dishes
- 1 scalpel or utility knife
- 1 forceps or tweezers
- 1 beaker of water

For the class:

- 4 plant pots 5–8 cm (2–3 in)
- potting soil
- 1 pkg radish seeds
- jars/beakers to cover plants and pots
- trays
- chart paper
- 2 boiling water baths
- 100 mL hot 95% ethyl alcohol
- 1 beaker (1000-mL) filled with tap water
- 100 g sodium hydroxide (flakes or pellets)
- medicine cups
- 1 beaker (1000-mL) for waste iodine

SAFETY NOTE: Always wear safety goggles when conducting experiments.

CAUTION: Sodium hydroxide can burn your skin. If it comes in contact with skin or clothing, rinse well with water.

NOTE: A chemical, sodium hydroxide (NaOH), absorbs the CO_2 gas in the air. By growing a plant in a closed system in the presence of NaOH, most of the CO_2 can be removed from the air.

PROCEDURE

1. **STOP & THINK** In the beginning of this module, you or your teacher germinated radish seeds under the following conditions:
   - in the presence of light, air, soil, and water
   - in the presence of air, soil, and water; in the absence of light
   - in the presence of light, air, soil, and water; in the absence of CO_2 in the air (in a closed environment)
   - in the presence of light, air, soil, and water (in a closed environment)

   In your notebook define the questions being asked in this experiment and your hypothesis.
2. Compare the characteristics of the plants in each pot, including height, color, ability to grow, and any other characteristics you may notice.
3. Label each of four petri dishes for one of the following: light, dark, closed normal, or closed no CO_2.
4. Take one leaf from a plant in each pot.
5. Mark each leaf you removed by notching it with a scalpel or knife (for example, no notches for those grown in normal conditions, one notch for those grown in the absence of CO_2, etc.). Be sure to write in your notebook how you marked each leaf (see Figure 3.3).

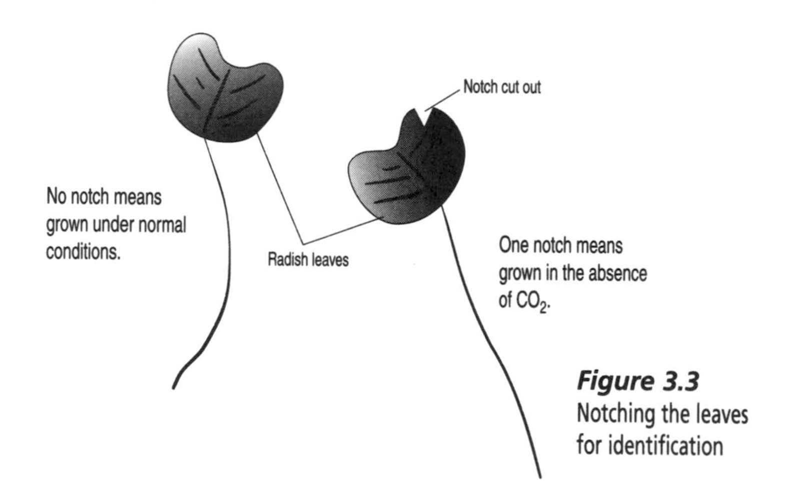

Figure 3.3
Notching the leaves for identification

CAUTION: Boiling water can cause burns. Ethyl alcohol can catch fire.

6. Immerse each leaf in boiling water for approximately one minute. Remove from boiling water with forceps.
7. Remove the pigments (chlorophyll and others) by immersing each leaf in hot 95% ethyl alcohol.
8. When most of the pigment (color) has been removed, use forceps to remove the leaf from the hot alcohol; dip each leaf in hot water again for a few seconds. This will keep the leaf from becoming brittle.

9. Place each leaf in the proper petri dish and cover with iodine solution. Allow leaves to sit for a few minutes. Iodine is used to detect the presence of starch, a carbohydrate that is an essential substance found in all living things. Iodine changes from brown to black or blue-black in the presence of starch.

CAUTION: Iodine stains and is a poison. Be sure to clean any spills thoroughly and avoid contact with skin and clothing.

10. Remove leaf with forceps and dip it into the beaker of clear water to rinse it. Pour iodine into the designated iodine waste beaker. Rinse out the petri dish to remove any remaining iodine solution.
11. Place each leaf back in the petri dish.
12. Compare the staining of each leaf and record in words and drawings what you observe.
13. Write a lab report which includes the following:
    - the questions being asked
    - your hypothesis
    - the experimental design, or how the experiment was set up
    - the data or observations you made
    - analysis and conclusions (your responses to the Analysis questions that follow)

    You may wish to discuss with your group the responses to the Analysis questions.

ANALYSIS

1. In an experimental design, the condition to be tested is a *variable.* That is, every condition in the experiment is held the same, or constant, except one—the variable. What was the variable in each of the experiments in the "You Light Up My Life" investigation? What were the conditions that were held constant?
2. The experiment in which the effect of CO_2 was measured needed to be designed as a closed system. What is meant by a "closed system"? What would have happened if the system had not been closed?
3. In addition to having only one variable, a good experiment will have a *positive control* or samples in which the outcome is known because there is no variable. Which plants served as your positive control? That is, which plants had all the appropriate resources and could be considered as having been grown under the best conditions regarding resources?
4. What do you think would eventually happen to the plants growing in the dark? to those growing without carbon dioxide? State your reasons.
5. Based on the experimental design, describe how you know starch is a product of photosynthesis. What would happen to those plants that were unable to synthesize starch?

6. On the basis of this activity, list the resources that a plant requires and speculate as to what each resource provides the plant and how the plant uses each resource.
7. When the interaction of two or more substances results in the formation of a different substance or substances, it is called a *chemical reaction*. The new substances have different physical and chemical properties than the starting substances. Chemical reactions are often expressed as an equation. For example, the reaction that results in water can be expressed as the joining of two components to form a third:

   oxygen + hydrogen —> water

   On the basis of what you have observed in this activity, write a word equation to describe what you think happens when plants are grown under appropriate environmental conditions with all the necessary resources for photosynthesis.
8. Assume two humans are provided with the same food, air, and water, but one person is raised in a part of the world where little light is present, while the other is raised in a sunny climate. Do you think their rates of growth will be affected by how much sunlight they get? Why or why not?

How Does Your Garden Grow?

The question of how plants get what they need to survive is one that has intrigued individuals for centuries. Speculations about plants and their feeding habits have given rise to many stories, such as the following:

> *On the island of Madagascar, off the east coast of Africa, the natives tell of a strange tree. This tree is an eater of meat, so they say—human meat. The person who ventures too close to the tree is seized by the tree's long branches and imprisoned. The branches wrap themselves so tightly around the victim that no matter how hard he or she struggles, the trap holds fast. Then slowly, but with great strength, the tree pulls its victim into its hollow center. There the body is digested, except for the bones, and the tree is nourished until another victim comes along.*
>
> *—From* Plants That Eat Insects: A Look at Carnivorous Plants *by Anabel Dean, Lerner Publications, Minneapolis, 1977, p. 7.*

In the following article, Isaac Asimov, scientist and writer of science fiction, asks this question in greater detail and presents one possible hypothesis based on an experiment carried out in the 1600s.

Plants somehow supply the food. It must, somehow, come from somewhere. It can't really form "out of nothing." [Plants require soil, water, air, and sunlight to grow. How does a plant take these substances and use them?] . . .

The man who had the thought [to find out] was Jan Baptiste van Helmont, an alchemist and physician of the Low Countries, who lived and worked in territory that is now Belgium, but was then part of the Spanish monarchy.

Van Helmont had the notion that water was the fundamental substance of the universe (as, in fact, certain ancient Greek philosophers had maintained). If so, everything was really water, and substances that didn't look like water were nevertheless water that had merely changed its form in some fashion. For instance, water was necessary to plant life. Could it be then, that, unlikely as it might seem on the surface, plant tissue was formed out of water, rather than out of soil? Why not try and see?

In 1648, van Helmont concluded his great experiment, great not only because it produced interesting and even crucial results, but because it was the first quantitative experiment ever conducted that involved a living organism. It was the first biological experiment...in which substances were weighed accurately and the carefully noted changes in weights supplied the answer being sought.

Van Helmont had begun by transplanting a shoot of a young willow tree into a large bucket of soil. He weighed the willow tree and the soil separately. Now if the willow tree formed its tissues by absorbing substances from the soil, then, as the willow tree gained weight, the soil would lose weight. Van Helmont carefully kept the soil covered so that no materials could fall into the bucket and confuse the manner in which the soil lost that weight.

Naturally, van Helmont had to water the willow tree; it wouldn't grow otherwise [see Figure 3.4].

Figure 3.4
Van Helmont's tree

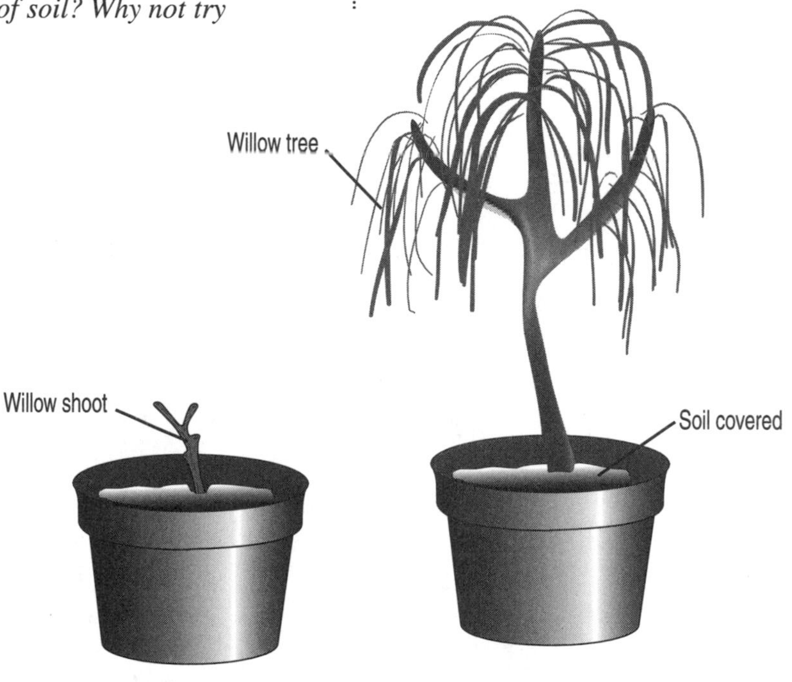

For five years, van Helmont watered his tree with rainwater. It grew and flourished and at the end of the time, he carefully removed it from the bucket, knocked the soil from its roots and weighed it. In five years of growing, the willow tree had added 164 pounds to its weight.

Very good! Now to weigh the soil after it had been dried. Had it lost 164 pounds to the tree? Not at all. It had lost only two ounces!

The willow tree had gained a great deal of weight—but not from the soil. What was the only other substance that made contact with the willow tree, van Helmont asked himself. The answer was: Water.

From this, he deduced that it was from water that the plant drew its substance, not from the soil. He used the results of this experiment to argue that water was indeed the fundamental substance of the universe, since if it could change to plant tissue it could surely change to anything else as well.

ANALYSIS

Write your responses to the following questions in your notebook.

1. What question did van Helmont set out to investigate in his experiment? What was his variable?
2. Was van Helmont's experimental design a good one, based on what was known at the time? Why or why not?
3. What conclusion did van Helmont reach about what constituted food for plants? Was his conclusion valid based upon the data he collected? Why or why not?
4. How do van Helmont's conclusions compare to the conclusions you reached in your plant investigation? Are they the same, different, or unrelated? Explain.
5. Based on your understanding of photosynthesis, redesign van Helmont's experiment and predict the results.

SUNLIGHT BECOMES YOU

What does sunlight provide for a plant? How does a plant use CO_2? What is "starch" and why does the plant need it?

The sunlight and air that a plant needs are used in the process of photosynthesis to generate *food* (substances which provide building blocks and energy for organisms) for a plant. An example of this food is the starch that you measured in the investigation "You Light Up My Life." The process of photosynthesis can actually be separated into two related biochemical processes: one which is dependent on the presence of light, and one which can occur in the absence of light (see Figure 3.5). During the light-dependent reaction energy in sunlight (solar energy) is absorbed by the green pigment that gives plants their characteristic color, chlorophyll. The chlorophyll, in turn, transfers this solar energy to another molecule, which stores it as chemical energy.

The chemical energy obtained during the light-dependent part of photosynthesis (called the "light reaction") is then used in the light-independent process (called the "dark reaction"). During this process, the atoms in carbon dioxide and water are rearranged to form new molecules of sugar, the substance that makes up starch. The sugar molecule contains carbon, oxygen, and hydrogen atoms, all of which were originally found in the carbon dioxide and water. In addition to sugar, this reaction also generates water and oxygen, formed from the hydrogen and oxygen atoms of the original water and carbon dioxide. The sugar molecule also contains chemical energy in the bonds between the atoms. Thus, the sugar molecule becomes the primary source of both energy and building materials for the plant.

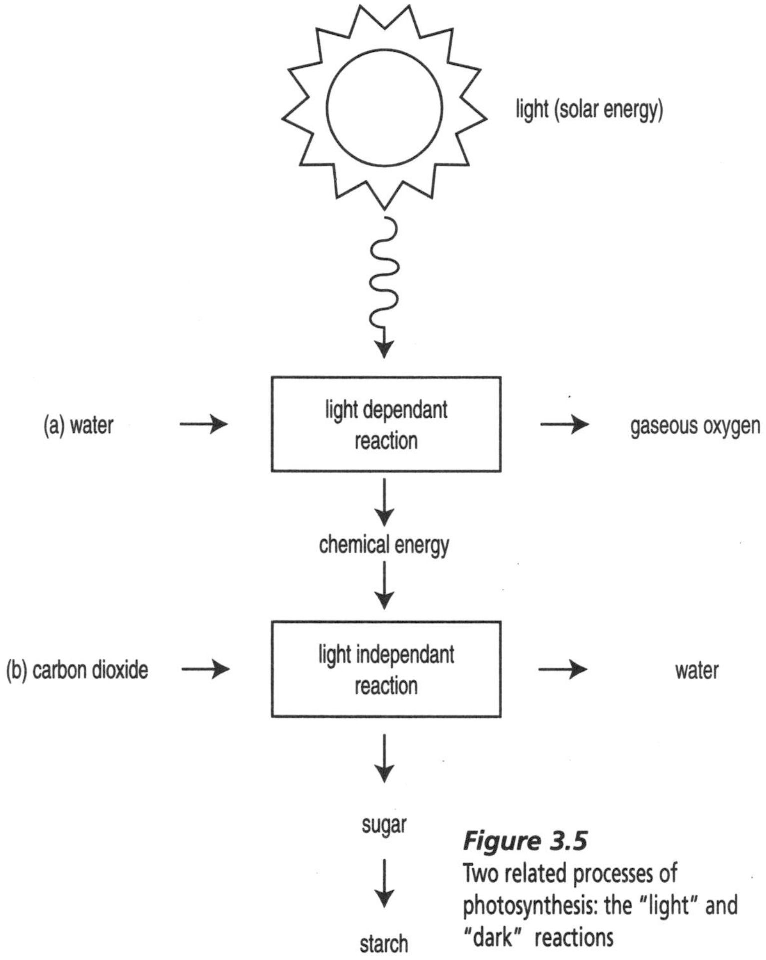

The word equation that you created during the analysis of your investigation can now be converted into a chemical equation:

light (solar energy)
(a) water → light dependant reaction → gaseous oxygen
chemical energy
(b) carbon dioxide → light independant reaction → water
sugar
starch

Figure 3.5
Two related processes of photosynthesis: the "light" and "dark" reactions

$$12\ H_20 + 6\ CO_2 + \text{energy} \longrightarrow C_6H_{12}0_6 + 6\ H_2O + 60_2$$

Plants require energy and building materials in order to carry out the processes of living. By the process of photosynthesis, plants obtain the nutrients and energy to grow, to respond to their environment, and to repair and maintain themselves. Without any one of the resources used in photosynthesis—air, water, sunlight—plants will "starve" to death. What happens to these molecules of $C_6H_{12}O_6$ sugar in the plant? Some of them remain as simple sugars such as fructose or glucose. Glucose, or grape sugar, gives that fruit its distinctly sugary taste. Another abundant sugar in plants is sucrose, also called table sugar or cane sugar because of its high concentration in the stems (canes) of plants, such as the sugar cane plant. Sucrose is synthesized when fructose and glucose are joined together to form a molecule containing two separate sugar molecules *(disaccharide)*. Sucrose is the major sugar that is transported throughout a plant and is the starting material for many other molecules in the plant.

Sugar molecules can also join together as links in long chains called *polysaccharides*. The starch that you tested for in the "You Light Up My Life" investigation is an example of a polysaccharide. Starch consists of many glucose molecules joined together. Starch serves as food storage for the plant; when food is needed, the starch is broken down into simple sugars. The sugars can then be transported wherever in the plant they are needed as building blocks and energy sources.

Some sugars form the starting materials for other large molecules in the plant. Living things are made up of several large *biomolecules* in addition to sugars (carbohydrates); these include proteins, lipids, nucleic acids, and fatty acids. In the next several learning experiences, you will explore what these biomolecules look like and how they are formed in living organisms.

The initial products of photosynthesis, sugar and starch, are produced in the cells which make up a plant's leaves. As described in the beginning of this learning experience (see Figures 3.1a and 3.1b), carbon dioxide from the air enters the leaf through tiny holes, or pores, called *stomata*. The stomata also allow release of one of the byproducts of photosynthesis, oxygen, back to the environment. In the leaf cells surrounding the stomata, the interaction of carbon dioxide, water (brought to the leaf through the stem), and light energy (absorbed by the green pigment chlorophyll) forms sugars which are transported through phloem in the stem to parts of the plant where they are needed. Some of the sugar is stored in the cells where it is converted into starch as a stored source of energy and building blocks.

Between the photosynthetic activities of the leaf and the absorbing activities of the root (which bring in minerals needed by the plant from the soil), all the nutritional needs of the plant are met. And by using these processes, plants can meet most of the nutritional needs of the rest of the living world.

ANALYSIS

1. Concept maps have a number of uses. One way to use a concept map is in the analysis of reading materials. Create a concept map

from the above reading that shows how photosynthesis occurs and how its products enable a plant to survive and maintain the characteristics of life. Include the following terms: photosynthesis, plants, sunlight, energy, chlorophyll, carbon dioxide, water, sugar, starch, leaf, stem, oxygen, root, biomolecules. (Use additional terms from the reading as you need them.) Place each term in a circle and join the circles with linking words.

2. Add to your map how the plant and other organisms might use the products of photosynthesis.

EXTENDING IDEAS

- Hydroponics is a method for growing plants and crops in the absence of soil. Research hydroponics and describe how this method is able to supply all the resources that plants need. Describe why some agriculturists might choose this method over more conventional methods.
- If you have access to the Logal software program *Biology Explorer: Photosynthesis*, design and conduct your own extension of van Helmont's research.
- During the Vietnam War, defoliating agents were used as a form of warfare. These agents were designed to destroy plant life in areas where they were used. Research the nature of the defoliating agents and how they worked. Based on your understanding of the role of photosynthesis in life, explain why this form of warfare was considered effective by some, inhumane by others.
- The Wetlands Protection Act limits actions such as cutting down trees in wetlands areas. Describe what would happen to the populations of smaller plants living under trees that were removed, and explain your predictions. If you have access to the Logal software program *Biology Explorer: Photosynthesis*, use the Independent Exploration: "Sun and Shade Plants" to help you investigate the light requirements of different plants. Similarly, the Independent Exploration "Delicate Balance" (which focuses on different levels of water availability and humidity) may be helpful.
- Changing climatic conditions alter the growth of plants. Explain how higher temperatures and greater concentrations of carbon dioxide influence the growth of plants and oxygen levels in the atmosphere. If you have access to the Logal software program *Biology Explorer: Photosynthesis*, use the Independent Exploration: "Origin of Fossil Fuels" to help you investigate the role of climate in plant growth.

LANDSCAPER AND GROUNDS MANAGER Do you like the outdoors? Landscapers and grounds managers who plan, design and maintain gardens, parks or lawns work in a variety of places—parks, schools, arboretums, zoos, botanical gardens, golf courses, and private homes. Depending on the job, they might be asked to dig and rake the grounds, prune trees, supervise the care of plants, trees and shrubs, decide if it is appropriate to use fertilizers or chemicals to help sick plants or trees, or identify potential pest problems. They might also decide on the types of plants, trees or shrubs that would be fitting in a specific setting, using environmental requirements such as average annual temperatures and precipitation, amount of sunlight available, and soil composition. A high school diploma is necessary for most positions; for a supervisory job, some college training or a college degree is necessary. Classes in subjects such as horticulture, botany, English, chemistry, business (accounting or economics), communications, plant pathology or landscape maintenance and design are suggested.

ORNAMENTAL HORTICULTURIST Have you ever thought you could work with nature to create a beautiful garden setting? Ornamental horticulturists raise and care for flowers, plants, trees, and shrubs in greenhouses and the outdoors. Horticulturists specialize in caring for fruits, nuts, berries, vegetables, flowers, and trees. People in these jobs place plants as ground cover in parks, playgrounds, or alongside highways. Jobs in horticulture include specializing in flowers or in trees, or in other plants commonly found in garden nurseries. To raise and care for flowers, they might also plant seeds or transplant seedlings, inspect crops for nutrient deficiencies or the presence of insects, identify diseases, determine the correct soil conditions, recommend which fertilizer to use to promote growth, or understand the environmentally controlled conditions in a greenhouse. These types of jobs might be found in florist shops, garden nurseries, or in research facilities. At a research facility, horticulturists might experiment in breeding, production, storage, or processing. Depending on the job, at least a high school diploma is necessary, but in most cases an associate's or bachelor's degree is preferred. Advanced degrees are necessary for horticulturists wanting to work in research facilities. Suggested classes include mathematics (algebra and geometry), chemistry, biology, English, horticultural design, and agriculture or botany.

FEEDING FRENZY

PROLOGUE **W**hy do we eat? Why do we eat what we eat? Using only air, sunlight, water, and a dash of minerals and vitamins, plants and other photosynthetic organisms can obtain all of the energy and manufacture all the materials they need to maintain the characteristics of life. Animals, however, are not so independent. They depend on plants to supply them with many of the resources required to sustain life. Animals may eat other animals, but every food chain originates with plants.

The sugars and other carbohydrates that plants synthesize serve as the source of energy and building blocks (that is, food) for the plant. Organisms that do not carry out photosynthesis must obtain all their nutritional needs by eating photosynthesizing organisms (plants) and other organisms in order to obtain the building blocks and energy necessary to maintain the characteristics of life.

In the words of Isaac Asimov:

> *If animals are to stay alive, then, they must find some source of food which doesn't have to eat, but which can produce its tissue substances seemingly "out of nothing."*
>
> *This would seem an impossibility (if we didn't know the answer in advance) but it isn't. The answer is plant life. All animals eat plants, or other animals that have eaten plants, or other animals that have eaten animals that have eaten plants, and so on. In the end, it all comes back to plants.*
>
> *Excerpted from* Photosynthesis, *by Isaac Asimov. Copyright (c) 1968 by Isaac Asimov. Reprinted by permission of BasicBooks, a division of HarperCollins Publishers, Inc.*

Imagine that aliens from another planet exploded a bomb containing an herbicide on the surface of the earth, extinguishing all forms of plant life. Write a news article or create a series of drawings that explain the consequences of such an event to animal life on Earth.

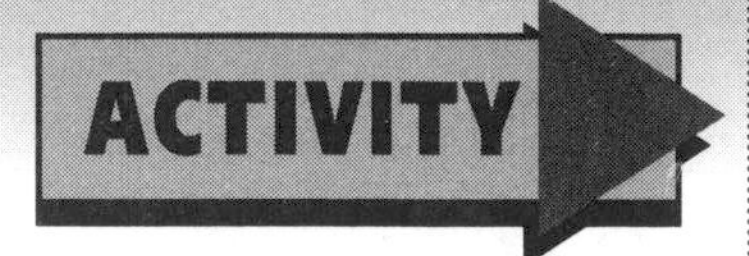

What Am I Eating, Anyway?

INTRODUCTION Around the world, different cultures have developed an assortment of diets which reflect their agricultural conditions, customs, and tastes. As different as they may seem at first glance, most of these diets supply the same nutritional requirements needed by humans to sustain life. What are these nutritional requirements and how can such different food sources as beef, rice, beans, insects, and vegetables all supply them?

How can we determine whether foods that seem so different in appearance are actually made up of the same or different components? In this investigation, you will identify some of the components of food which are required to sustain life. These components which make up food are called *nutrients*. A nutrient is considered "essential" when it cannot be synthesized by the organism but must be obtained by the organism from its environment.

You will be using *indicators* as chemical detection tools to find out what nutrients are present in foods. Indicators are chemical compounds used to detect the presence of other compounds. Detection is based upon observing a chemical change that is taking place. The substance being detected and the indicator together are involved in a chemical reaction which brings an observable change—most often, a change in color.

MATERIALS NEEDED

For each group of four students:

- 4 safety goggles
- 4 pairs of disposable gloves (optional)
- 20 test tubes
- 1 test tube clamp
- 1 test tube rack
- 1 clean eyedropper
- 1 glass stirring rod
- distilled water in a beaker
- 1 wax marking pencil
- 3 food samples (or more if available)
- 3 beakers (250-mL)
- 1 small bottle Benedict's solution
- 1 small bottle iodine
- 1 small bottle Biuret reagent
- 1 small bottle Sudan III or IV reagent (or brown paper)
- forks
- plastic containers or mortar and pestle
- 50-mL graduated cylinder (optional)

For the class:

- positive test controls for:
  - sugar
  - starch
  - protein
  - lipid
- boiling water bath
- blender (optional)

PRE-LABORATORY ANALYSIS

Read the laboratory procedure carefully and write responses to the following questions in your notebook.

1. How would you pursue finding out which nutrients foods have in common?
2. What do indicators tell us?
3. Before you do the experiment, in what ways are you treating the food? Why?
4. How will you know that each indicator has worked?
5. What are some ways that you can record data and keep track of which indicators are giving a positive result for which nutrients?

PROCEDURE

SAFETY NOTE: *Always wear safety goggles when conducting experiments.*

1. STOP & THINK Collect your group's food samples. Predict which nutrients are present in each sample you are about to test. Record your predictions in your notebook.
2. Set up four test tubes. With a wax marking pencil, label all four tubes with a minus sign (to indicate negative control) and then label each with one of the indicator solutions, as follows:

   - sugar - protein
   - starch - lipid

   After labeling the test tubes, place them in a test tube rack.

   Using an eyedropper, place 15 drops of each indicator (from Table 4.1 in the margin) in the appropriately labeled test tube.

   Be sure to rinse the eyedropper between solutions to reduce contamination. To rinse: draw distilled water from one beaker into the dropper, then squirt the water into a second beaker. Repeat 2–3 times for a thorough rinse.

CAUTION: *Biuret solution can burn your skin. If it comes in contact with skin or clothing, rinse thoroughly with water.*

Table 4.1
Key for Nutrient indicators

NUTRIENT	SOLUTION
sugar	Benedict's solution
starch	iodine solution
protein	Biuret solution
lipid	Sudan III (or IV) solution

3. Add 15 drops of water to each test tube. These are the negative test controls.
4. Obtain three more sets of four test tubes (to identify the food nutrients in each food sample). Label the set 1 tubes as follows:

   1/sugar, 1/starch, 1/protein, 1/lipid.

   Label set 2 as follows:

   2/sugar, 2/starch, 2/protein, 2/lipid.

   And so on. As you label the tubes, place them in your rack.
5. Obtain four more test tubes and label them to indicate positive controls:

+ sugar + protein
+ starch + lipid

Place them in your rack. Be sure to have all the same indicators in one row, starting with the negative control at one end and ending with the positive control at the other end of the row, with the samples to be tested in between. (See Figure 4.2.)

6. Write down the name of each sample in your notebook, and assign each sample a number from 1 through 3. Copy Figure 4.2 into your notebook.

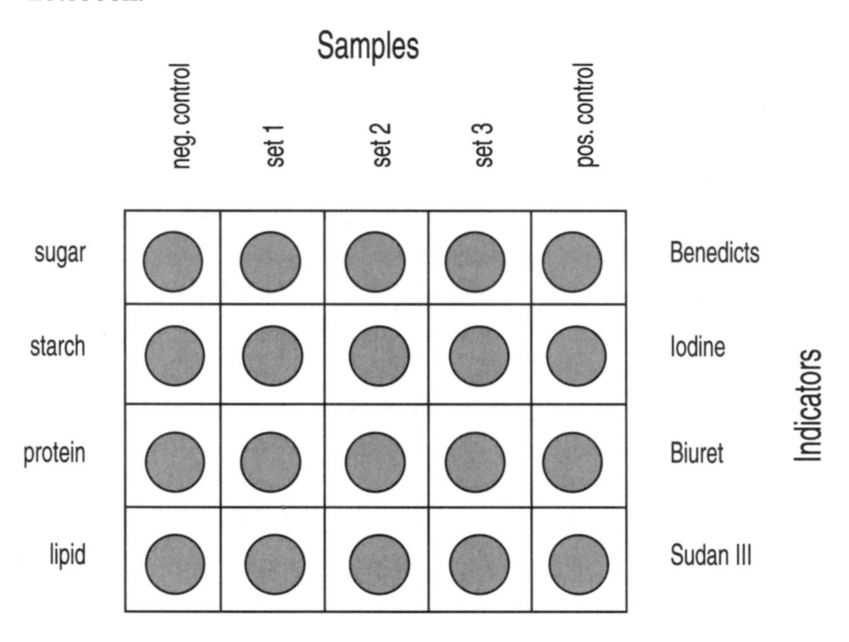

Figure 4.2
Test tube setup for nutrient analysis tests.

7. Take food sample 1 and mash, blend, or dissolve it in enough water to make it liquid. Use the glass rod to mash or crush the food (or blend it in a blender), and add about 10–20 drops of water, then stir with the glass rod. Place small amounts into each of the set 1 test tubes. Repeat with your second and third food samples, adding them to set 2 and set 3 test tubes respectively.

8. Obtain from your teacher the positive test controls—sugar, starch, protein, and lipid. Add 15 drops of each to the appropriately labeled tubes. Be sure that, if you are using the same eyedropper, you rinse it thoroughly in water between taking samples. The positive controls are your known standards. Because you know which nutrient each is composed of, the controls can be used to determine what a positive test for each indicator looks like. Because chemical changes in indicator testing are not always obvious, it is important to become familiar with exactly what a positive test looks like, so that you can be informed and experienced when you begin testing unknowns.

NOTE: Known standards give you an opportunity to make sure indicators are working properly. Obtaining negative results with standard solutions can alert you to possible problems with the indicators.

9. **STOP & THINK** Why is it important that you prepare your negative controls first and your positive controls last? Record your response in your notebook.

10. How to perform indicator tests:

    When adding your indicators to each tube be careful not to touch the side of the test tube with the eyedropper. Hold the dropper over the tube mouth and let the drops "free fall" into the tube. After adding the indicators, give each tube a gentle tap with your forefinger to mix.

    Add 15 drops of Benedict's solution to each of the three tubes labeled "sugar." Heat the tubes in a boiling water bath for five minutes without disturbing. (You may continue on to other tests while waiting for this test.) Use a test tube clamp to lift test tubes out of the boiling water bath in order to examine the contents.

    Benedict's indicates the presence of sugars (or simple carbohydrates). Compare your results with the positive and negative test controls. Describe the results in your notebook.

11. Add 15 drops of iodine solution to each of the three test tubes labeled "starch." Examine the contents of each tube. Compare your results with the positive and negative test controls. Describe the results in your notebook. Is it positive or negative for that nutrient?

12. Add 15 drops of Biuret solution to each of the three test tubes labeled "protein." Examine the contents of each tube. Compare your results with the positive and negative test controls. Describe the results in your notebook.

13. Add 5 drops of Sudan III or IV to each of the three test tubes labeled "lipid." Examine the contents of each tube. Sudan III and IV indicate the presence of lipids (fats). Compare your results with the positive and negative test controls. Describe the results in your notebook.

14. Write a laboratory report which includes the following information:
    - the question being asked and your predictions about experimental results
    - a brief description of the procedure (including how indicators work)
    - a data table with the results of your experiment and the observations you recorded
    - discussion (include in this section your responses to the analysis questions below)
    - your conclusions (do not repeat data results)

Figure 4.3
Removing test-tubes from a boiling water bath

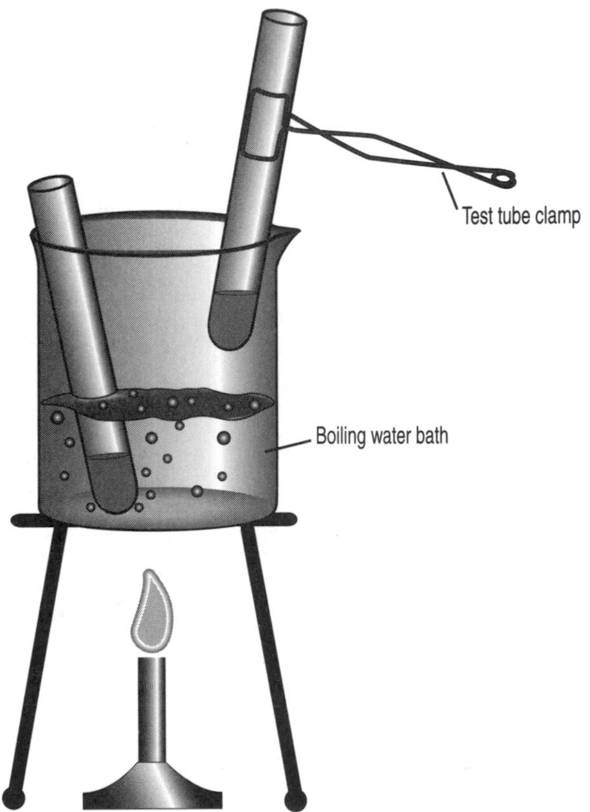

CAUTION: *Hot plates and boiling water can cause burns. Handle the test tube clamp carefully to avoid dropping the test tube or splashing the water.*

NOTE: You may need to hold the test tubes up to a white sheet of paper for easier observation.

NOTE: Be sure to rinse the eyedropper between each solution.

ANALYSIS

1. What was the difference between a positive result and a negative result?
2. Based on your observations, what function do negative controls serve?
3. What function do positive controls serve?
4. Why was it important to rinse your eyedropper between tests?
5. Describe any sources of error which might have affected your results.
6. What do you think the nutrients that you identified in these foods supply to you?
7. Through reading and investigating, you have identified certain nutrients which are present in foods. Create a table listing these nutrients across the top. List the foods you have eaten today along the left side of the table. Based on the results from the class data about which foods contain which nutrients (sugar, starch, protein, lipid), determine which nutrient(s) you received from each of the foods you ate by placing a check in the appropriate column. What do you think the consequences of receiving inadequate amounts of one of these nutrients might be?
8. What has surprised you about your own diet?

The Missing Ingredients

Why are we encouraged by parents, teachers, and the U.S. government to eat well-balanced meals? What are these foods providing us? If an organism is to sustain life it must be able to obtain building blocks and energy from the biomolecules making up its food. As you have determined, non-photosynthesizing organisms must obtain from their environment three basic types of nutrients: proteins, carbohydrates, and fats. In addition, they require certain amounts of vitamins and minerals. The following is an account of what happened when an organism, in this case a human, does not get an adequate balance and amount of nutrients.

MADELYN'S STORY

When I was 12, I put myself on a diet. I would drink one glass of powdered skim milk and eat one hard-boiled egg. I would eat an egg because that was a measured amount, and I used to agonize over the size of the egg after awhile. I stayed on that diet for a month and a half—it was my secret. I got a baby-

sitting job after school just so that I would not have to eat with my family. If I was not baby-sitting, I was at the library.

I was up all night exercising. I would wear layers of clothes so no one had any idea of what I was doing. If anyone said I looked drawn, haggard, or pale I would redouble my attempt to appear normal, all the time keeping to my 150-Calorie-a-day diet. The next 20 years I spent trying to follow that diet and regain the feeling of control I had while on it.

One night I was dragged to a relative's house for dinner. I remember eating a piece of lettuce and feeling that the whole month and a half was for nothing, that my diet had been ruined. I wanted to kill myself. I felt very out of control. Later I realized that I thought then that my body and what I ate was the only thing I could control, and I was very good at that. As a typical perfectionist I did not want to do what I was not good at . . .

—An excerpt from Eating Habits and Disorders *by Rachel Epstein, New York, Chelsea House Publishers, 1990*

Madelyn was in the process of self-starvation. In drinking only one glass of skim milk and eating one hard-boiled egg, she severely limited her nutrient and Calorie uptake.

What does a person's body do when placed under this form of stress? The body has no way of storing proteins. Even after just one day without protein, the body will begin to break down the protein in nonessential tissues, such as muscles, and "recycle" the components for essential functions. Eventually the body will begin to break down the protein in essential tissues, like the heart or the liver, and serious health problems result.

Carbohydrates are the major source of energy for the body. Studies have indicated that in a healthy diet, 60% of the energy we use should come from carbohydrates. The energy comes from the supply of glucose that the organism has stored as carbohydrate. In plants, the chief energy storage molecule is *starch*; many organisms, including humans, use plant starch as a primary source of food. In humans, glucose is stored as *glycogen* in muscle tissue; many long distance runners "carbohydrate load" before a long race by eating enormous quantities of pasta which provide them with a ready source of energy. Carbohydrates also are important in forming structural components of organisms.

Fats form a fundamental part of an organism's structure and function and are as essential to maintaining life as proteins and carbohydrates. Fats are also a major energy storage form, but the energy in fats is not as readily available as that of carbohydrates. Fats are a member of a class of nutrients called lipids; hormones, cholesterol, and some vitamins are also members of the lipid family of molecules. *Lipids* form essential structural components of membranes, store and circulate certain vitamins, and store energy in reserve. Without an intake of a certain amount of fat, organisms cannot maintain these important functions.

> *Weight was the whole focus of my life. I [Madelyn] had been aware of it even before I began dieting. My older sister was very overweight and usually left the dinner table in tears. I felt very guilty about it, but I remember thinking I was glad I had an older sister like her so I would know I should never become that fat, that it was too painful. My older sister had no friends and was ridiculed by my father (*Epstein, 1990*).*

The long-term consequences of Madelyn's disorder (anorexia) are frightening. Madelyn's symptoms show what can happen if appropriate nutrients are missing:

- generalized fatigue, lethargy and lack of energy
- paleness or grayish tones to skin, becoming dark and scaly
- cramping of muscles—muscles eventually waste away, making all physical activity difficult
- numbness or tingling sensations in the hands or feet
- stomach bloating, constipation, and difficulty urinating; kidney and bladder infections
- periods of dizziness, light-headedness, and even amnesia
- shrinkage of internal organs, which may be irreversible in extreme cases
- kidney failure
- heart failure

Food plays an important role not only in maintaining health but in our social lives. We entertain with food, we signify certain events with food, we use food symbolically in cultural and religious events, we reward with food. In many societies, food is also associated with aiding performance and personal appearance. We eat certain things to do well at sports or to think better (weight lifters eat protein; fish is known as "brain food"); we may or may not eat certain things hoping to enhance our appearance (thinking that eating gelatin strengthens nails; drinking water clears the complexion, chocolate does the opposite; cheesecake goes right to the hips and thighs). In some instances, this consciousness of what we "should or should not eat" can result in skewed and even dangerous eating habits.

Dieting is a hallmark of American culture. Books on how to take off unwanted pounds abound, diet programs are ubiquitous, and fat-free foods and products to "take it off fast" fill our supermarket and pharmacy shelves. Most of these "get-thin-quick" approaches are destined to fail for most people. Losing weight, in most cases, is a matter of a consistent plan of increasing physical activity to a level that exceeds the daily intake of food. When this happens, the glycogen reserves in liver and muscle are depleted. When these are gone, the body takes fats out of storage and uses them for energy. Continued on a regular basis, this approach results in a slow but steady reduction in fat reserves and weight loss. For some the solution is not so simple; metabolic rates and genetic heritage may override even the most dedicated dieter. But for most people a well-balanced diet coupled with daily activity will ensure that the needs of the body will be met as weight is lost.

ANALYSIS

Using the reading and any other information you may have, respond to the following questions in your notebook:

1. This article describes some of the symptoms of anorexia nervosa. Describe what nutrients were lacking in Madelyn's diet and how this relates to the symptoms described.
2. What factors might contribute to a person severely limiting his or her intake of adequate amounts of nutritional substances?
3. How might severe reductions in a person's diet start to affect the body? Look back at "Madelyn's Story." Could someone cut out all the fat in his or her diet, lose weight, and still be healthy? All the protein? All the carbohydrates? Explain why or why not.
4. The eating disorder described is severe; at one time or another many individuals limit their intake of certain foods for a variety of reasons without the serious consequences described. Yet limiting nutrients can have health consequences even when done in a less intensive fashion. Even if one is dieting to lose weight, nutrients should be obtained in a balanced fashion.
   a. Using the nutrient information from a variety of food labels, nutrition/diet books, or the Nutrient Content of Food Table found in Appendix A, create a menu based on the following intake for an average person (2,000 Calories for a female, 2,500 Calories for a male) to maintain his or her weight:

   Daily Recommended Intake for an adult wishing to maintain weight:

Fat	**60 grams total***
Protein	**55 grams total**
Carbohydrate	**250 grams total**
Calories	**2500 Calories**

(*Calories obtained from fat are not exactly comparable to Calories obtained from carbohydrates. For this analysis, you should assume they are comparable.)

b. Using the same information, reduce the total Calorie intake by about one-third (for a new total of 1,500 Calories). What could you remove or change in the original diet to keep the diet balanced? What amounts and types of foods would constitute a balanced diet while allowing the person to lose weight?

c. Take this one more step by reducing the above intake by about half (800 Calories total daily). Could a balanced diet be maintained at this level? (This is now approaching an anorexic diet.)

EXTENDING IDEAS

- Research the discovery of one of the nutrients. How was it identified? How did researchers determine its role in the body?
- The requirements for certain vitamins and minerals were determined by studying diseases that result from deficiencies. For example, the importance of vitamin C was determined by investigations which focused on scurvy, once a common disease of sailors. Research a disease caused by a vitamin or mineral deficiency; describe what the symptoms are and why they occur, based on your understanding of the role of the substance in the body.
- Certain countries concentrate their agricultural efforts on a single crop—rice. These are often countries in which not much meat is consumed because of economics or dietary restrictions. Describe why this could create a health problem. Design a possible solution and explore the economic, social, and health issues involved in such a solution.
- Research how the FDA (Food and Drug Administration) conducts its food testing. How are Calories measured? How are the amounts of protein, carbohydrates, fats, vitamins and minerals determined? How many batches of each food need to be measured before an average is taken? Why do you think the government is interested in the nutritional state of its citizenry?
- Research the concept of a limiting nutrient. How do limiting nutrients affect populations? What happens when the limiting nutrient is supplied in excess?
- Research various commercial diet plans. What is the total nutrient value of each? Are any essential nutrients missing? Describe why each of the diets researched would or would not be good for the body on a long-term basis.
- The diets of a group of individuals may be determined by factors other than what is best for them as living organisms. One factor in

the United States is advertising. Find newspaper and magazine advertisements which are designed to influence eating habits. Determine whether their claims are accurate based on your understanding of nutritional needs of humans, and describe why they are or are not appropriate.

ON THE JOB

NUTRITIONIST Would you be interested in a career where you think about food all day? Nutritionists use their science background to look at the relationship between human nutritional needs and what nutrients specific foods provide. They focus on research and education about food and its composition. If a nutritionist focuses on the biochemistry of food (that is, the sugar, lipid, starch, and protein components), the job might seem more like that of a food scientist. For someone having a nutritional biochemistry background, jobs are available in the government, in hospitals, and in research laboratories at universities or pharmaceutical companies. A nutritionist who focuses on educating the general public about health is communicating what good nutrition is. A nutritionist like this might work as a teacher in a school, in a medical practice talking individually with people, or as a writer for a health or nutrition column in a magazine. Some nutritionists are also registered dietitians. Thirdly, a nutritionist might focus on research and development in the food sector by contributing to new product development. This would involve developing new recipes or revising old recipes so that the products which appear on the grocery store shelves are lower in fat, lower in sodium, and still taste good. The training to become a nutritionist consists of a four-year college degree in a nutrition program, a master's degree in nutrition and a professional internship with a registered dietitian. Classes in subjects such as biology, chemistry, physics, English, and computers are useful.

DIETITIAN Would you be interested in a career where you work with people and think about food all day? Dietitians look at the relationships between food composition and how the human body uses the nutrients found in food. Dietitians, who can also be nutritionists, spend time talking with their clients and planning and interpreting an individual's daily nutritional needs. For instance, clients might visit a dietitian to get advice on nutritional issues, to learn how to change their diet because of a high cholesterol level or diabetes, or to find out how to lose weight while still getting all the nutrients they need. Dietitians and nutritionists might also work with a psychologist or psychiatrist to counsel people who have food disorders. Dietitians might work in private practice or in a hospital.

THE LEGO OF LIFE

PROLOGUE What happens to food in living organisms? You may have heard the phrase "You are what you eat." But what does what you eat have to do with what you are?

The diversity of foods is caused by different variations in the components that make up food, the building materials of food. In any construction work, the kind of building materials you need is determined by the kind of structure you want to build. So, it makes sense that foods—which are living, once-living, or at least made up in part of once-living organisms—should provide the right materials for building organisms. Each day of photosynthesis by a plant, or each meal eaten by an animal, is a shipment of supplies for that organism's construction work. But how does that happen?

In Learning Experience 4—Feeding Frenzy, your work with chemical indicators demonstrated that most foods are actually composites of several different components—nutrients—which include sugars and starches (carbohydrates), fats (lipids), and protein. Exactly what are these biomolecules? What are they made of? What do they contain that living organisms need?

In this learning experience, you examine the chemical composition of the biomolecules that make up food and compare their chemical components and structures.

MADAM, I'M ATOM

What happens to the hamburger and fries that you ate at lunch? What are the steps that occur in your body that enable you to use those delicacies in order to maintain the characteristics of life? In order to make them accessible to you, these foods must be broken down into forms your body can work with. Food taken into your body gets broken down into smaller and smaller components. Beginning in your mouth, food is digested or broken down as it passes through your digestive system either mechanically by teeth for example, or by chemicals designed to break down the biomolecules into their component parts.

What are these parts, and what are they made of? How do scientists find out what things are made of? What is the smallest unit into which things can be subdivided?

More than 2,000 years ago, people were already wrestling with the idea of what the smallest unit of matter might be. A Greek philosopher named Democritus thought that you could keep cutting a material into smaller and smaller pieces until finally you would get to a piece so small it was uncuttable. He thought that any kind of matter—like an apple—could be divided only so many times. Each cut would take him closer to the smallest piece, which itself would be uncuttable. He called this uncuttable piece an atom.

Democritus never actually saw an atom. But it made sense to him that all matter was made up of smaller parts. What he did not know was whether or not those smaller parts were all identical. In other words, was an apple made up of "atoms of apple" or did an apple contain atoms which were also found in other substances such as water, wood, and soil?

Aristotle, another Greek philosopher, had a different take on what made up something like an apple. He thought that everything was made of the same four basic things which he called elements: earth, water, air, and fire. How could this be? How can an apple be made of fire or air or soil? Aristotle might have explained this by saying that the apple tree roots carved into the soil, that an apple tree grows from the earth and that the tree gained substances and water in the earth. With a few simple experiments, Aristotle might have observed that without air, the tree could not have survived. Finally, Aristotle might have held a torch to the tree, and observed that the tree "gives off a lot of fire." Aristotle hypothesized that the apple tree must be made of earth, water, air, and fire.

Figure 5.1
Democritus
(The Granger Collection)

Van Helmont, whose experiments you studied in an earlier learning experience, had many ideas that were similar to Aristotle's. He thought that whatever was taken in from the earth made up the plant. When he found that the soil only lost two ounces while the plant gained dozens of pounds, he turned to water to explain the plant's growth. He thought that water must be what makes up the plant, even though water and the plant did not look at all alike. Neither Aristotle nor van Helmont considered that water, earth, and air might be composed of smaller pieces. They didn't think that earth, for example, might be made of smaller pieces which could be rearranged into something else.

In the late 1700s, a scientist named Antoine Lavoisier demonstrated what Aristotle and van Helmont could not imagine: that water could be divided into smaller pieces which were not themselves water. By applying enough energy to separate the component atoms, Lavoisier found that water was made up of two smaller components—hydrogen and oxygen. He could not separate hydrogen and oxygen into any smaller components, at least in ways he could detect. Lavoisier's experiment demonstrated that one of Aristotle's basic "elements," water, was composed of smaller elements which could join together to make a substance very different from the starting material. Another scientist, Joseph Priestley, dismantled another of Aristotle's "elements," air, into smaller pieces which included oxygen, a component of water.

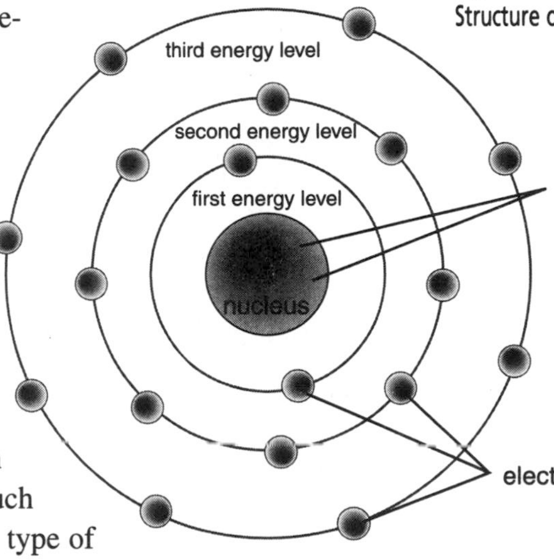

In 1808, John Dalton, an English meteorologist, formally presented the modern atomic theory, sort of a modern-day version of Democritus's theory. Dalton proposed that for each chemical element such as oxygen or hydrogen, there was a unique type of indivisible object called an atom. In fact, *elements* are defined as substances that are composed of only one kind of atom. When two or more atoms join together, they form a *molecule*. Molecules make up most of the different kinds of material in the world.

Figure 5.2
Structure of an atom

The concept of one atom, indivisible, came tumbling down in 1897 when the English physicist Joseph John Thomson identified a particle called an *electron*, which was smaller and lighter than any atom known. He concluded that atoms, themselves, were not the fundamental building block of matter but were made up of even smaller and more fundamental particles. In the early twentieth century scientists discovered two more subatomic particles, the *proton* and the *neutron*. These were heavier than electrons and were in the center, or *nucleus*, of the atom. The number of protons in the nucleus determines the type of atom; for example, all hydrogen atoms have one proton and all carbon atoms have six protons.

As scientists learned more about what matter was composed of, they began to learn that living matter and nonliving matter were very different. Nonliving matter, such as minerals in rocks, table salt, water, and air, can be made up of a wide variety of elements. However, living matter seems to be composed of a very small number of elements. Six of these elements—carbon, hydrogen, oxygen, nitrogen, sulfur and phosphorus—are present in great abundance and, as you will see, make up the major components of living organisms. Living organisms also contain other elements but in much smaller amounts. For example, plants actually contain 17 different elements, the six that are found most abundantly and eleven others.

Figure 5.3
Periodic table of the elements

Key: 1 = Atomic number; H = Symbol; 1.008 = Atomic mass

Transition Metals: groups 3–12

	1 1A	2 2A	3 3B	4 4B	5 5B	6 6B	7 7B	8 8B	9 8B	10 8B	11 1B	12 2B	13 3A	14 4A	15 5A	16 6A	17 7A	18 8A
1	1 H 1.008																	2 He 4.003
2	3 Li 6.941	4 Be 9.012											5 B 10.81	6 C 12.011	7 N 14.007	8 O 15.999	9 F 18.998	10 Ne 20.179
3	11 Na 22.990	12 Mg 24.305											13 Al 26.982	14 Si 28.086	15 P 30.974	16 S 32.06	17 Cl 35.453	18 Ar 39.948
4	19 K 39.098	20 Ca 40.08	21 Sc 44.956	22 Ti 47.90	23 V 50.942	24 Cr 51.996	25 Mn 54.938	26 Fe 55.847	27 Co 58.933	28 Ni 58.71	29 Cu 63.546	30 Zn 65.38	31 Ga 69.72	32 Ge 72.59	33 As 74.922	34 Se 78.96	35 Br 79.904	36 Kr 83.80
5	37 Rb 85.468	38 Sr 87.62	39 Y 88.906	40 Zr 91.22	41 Nb 92.906	42 Mo 95.94	43 Tc 98.906	44 Ru 101.07	45 Rh 102.906	46 Pd 106.4	47 Ag 107.868	48 Cd 112.40	49 In 114.82	50 Sn 118.69	51 Sb 121.75	52 Te 127.60	53 I 126.904	54 Xe 131.30
6	55 Cs 132.905	56 Ba 137.34	57 La 138.906	72 Hf 178.49	73 Ta 180.948	74 W 183.85	75 Re 186.2	76 Os 190.2	77 Ir 192.22	78 Pt 195.09	79 Au 196.966	80 Hg 200.59	81 Tl 204.37	82 Pb 207.19	83 Bi 208.980	84 Po (210)	85 At (210)	86 Rn (222)
7	87 Fr (223)	88 Ra (226.025)	89 Ac (227)	104 Unq (261)	105 Unp (262)	106 Unh (263)	107 Uns (262)	108 Uno (265)	109 Une (266)									

Inner Transition Metals

Lanthanide Series	58 Ce 140.12	59 Pr 140.908	60 Nd 144.24	61 Pm (147)	62 Sm 150.4	63 Eu 151.96	64 Gd 157.25	65 Tb 158.925	66 Dy 162.50	67 Ho 164.930	68 Er 167.26	69 Tm 168.934	70 Yb 173.04	71 Lu 174.97
Actinide Series	90 Th 232.038	91 Pa 231.036	92 U 238.029	93 Np 237.048	94 Pu (244)	95 Am (243)	96 Cm (247)	97 Bk (247)	98 Cf (251)	99 Es (254)	100 Fm (257)	101 Md (258)	102 No (255)	103 Lr (256)

Of the more than 100 elements known (see Figure 5.3), only 22 appear to be essential to life. This seems hard to believe when we observe the great diversity of living organisms around us. How could all living things be composed of just a few kinds of elements? What might be the reason?

ANALYSIS

1. Create a diagram that flows from largest to smallest, using the following substances; atoms, neutron, electron, carbohydrate, cow, sugar, element, proton, lipid, nucleic acid, protein, water, molecule,

corn, nucleus. You may use a word more than once. If you are not sure which of two items is larger, place them side by side in your flow diagram.

2. Describe any relationships that might exist among the various substances (for example, cow eats corn).

A Mere Six Ingredients

INTRODUCTION The fundamental driving force that determines what organisms take in from their environment is what they require in order to survive. What role do nutrients play in sustaining life, and how do organisms use nutrients?

All living organisms must obtain building blocks for making new biomolecules and energy to carry out the essential processes of life. These building blocks and energy are found in the biomolecules that make up food and that also make up all living things: carbohydrates, lipids, nucleic acids, and proteins. These nutrients in turn are made up of only six elements, which are used to build a tremendous diversity of living things by being arranged in different ways. In living organisms atoms of the six elements (carbon, hydrogen, oxygen, sulfur, phosphorus, and nitrogen) are joined in different arrangements and in different quantities to form all of the different biomolecules.

How do different organisms obtain these nutrients? For most organisms, there are only two possible ways: by constructing them during photosynthesis, or by feeding on other organisms. During photosynthesis, plants take in water and carbon dioxide from their environment. Using solar energy, the elements (hydrogen, oxygen, and carbon) in these molecules are "recycled" and used to make a new, energy-containing molecule: sugar. Since plants do not "eat" and, therefore, do not take in biomolecules such as protein and lipid, plants must use this sugar molecule to construct all the carbohydrates, lipids, and proteins they require to sustain life. The sugar, in conjunction with vitamins and minerals which the plant obtains from soil, provides all the essential nutrients a plant needs.

Organisms that do not carry on photosynthesis are required to take in complex biomolecules from their environment by eating plants (with the nutrients the plants have manufactured) or other animals (that have eaten plants or still other plant-eating animals). Ultimately, all the nutrients animals take in can be traced back to plants. Thus, the phrase "You are what you eat" is a very literal one. In plants and other photosynthetic organisms, food is manufactured by the process of photosynthesis. In animals, complex biomolecules are taken in as food and the elements which make them up are recycled into biomolecules which make up the animal. Six elements—carbon, hydrogen, oxygen, sulfur, phosphorus, and nitrogen—joined in different arrangements and in different quantities, are the main ingredients of life.

How do living organisms rearrange the components of their food? What do the biomolecules of life look like?

MATERIALS NEEDED

For each group of students:

- 1 set of molecular models (optional)

TASK

1. What elements make up the biomolecule you are examining?
2. What is the shape of the subunit for your biomolecule?
3. How do these subunits join together to form the biomolecule?
4. Show how the subunits can link together to form different kinds of biomolecules in your group. For example:
   - For carbohydrates, show how subunits are linked together to form disaccharides, starch, and cellulose;
   - For lipids, show the differences between saturated and unsaturated fats; show how these fats form phospholipids, or cholesterol;
   - For proteins, show how proteins can vary in composition;
   - For nucleic acids, describe different kinds of nucleic acids and indicate how nucleic acids can vary.
5. State several functions that your biomolecule fulfills in an organism.
6. If molecular models are available, build a three-dimensional model of your subunit using molecular models.

EXTENDING IDEAS

- Research the digestive system and describe or draw how nutrients are broken down in the various organs. Describe the digestive enzymes involved and how they act.
- Some vegetarians assert that obtaining protein from animal products is economically very inefficient and that greater amounts of protein can be obtained by consuming only plant products. Many in the

beef and dairy industries disagree. Research the arguments used by each side and decide which one you would support in a debate and explain why.

- During the Cultural Revolution in China in the 1960s, famine forced many people to try to supplement their diet by eating leaves from trees. In spite of that, many died. Cellulose and starch are both long-chain polysaccharides found in plants. Both are made up exclusively of glucose molecules. If the leaves of the trees were mainly cellulose, why could humans not survive on leaves? Describe and explain the difference between a human's ability to use starch and ability to use cellulose.

ON THE JOB

CHEMICAL TECHNICIAN Have you ever wondered why oil and water don't mix, but why you can dissolve sugar into water? Chemical technicians assist during the development, testing, and manufacturing of products in either a basic research laboratory or an applied research laboratory. Chemical technicians look at the chemical content of a product, its purity, strength or stability. Chemical technicians might specialize in one type of activity or in one specific type of product. During a research and development phase, a technician might be working on creating new chemical products or on new ways in which to make chemicals from different beginning materials. This might include taking measurements, making calculations, or collecting and analyzing data. During the design and production phase of making a product, a technician might plan the process to create a product and then design and operate the equipment used to carry out the process. Once a product is being manufactured, the role of a technician might be to ensure quality. During a quality control phase, the manufacturing process is watched and the final product is tested to ensure that it meets the required specifications, which might include environmental standards. Once the company is manufacturing the product and it is available on the market, a chemical technician might also be a customer service representative. It is useful for representatives who work with existing business customers and who try to recruit new customers to understand the nature of the chemical products and processes of their own company's products, as well as those of the customer. Chemical technicians are found not only in the chemical industry, but also working in the food, petroleum, aerospace, electronics, paper, automotive, or construction industries. A minimum of a two year college degree is required. Classes in algebra, geometry, sciences (preferably chemistry), English, and computers are recommended.

HISTORIAN Did you know that Edmund Halley, who first predicted how long it would be before the comet now known as Halley's comet would again be visible to Earth, also worked with Isaac Newton? Like other fields, science has a long history. Historians can specialize in science, technology and medicine as easily as they can specialize in American history. Historians of science look at the interactions of science, technology, and medicine with the cultural, social, political, and economic forces of the time period. With knowledge about the history of science, an individual might work in a museum developing exhibits (choosing what to put on display and writing labels for the general public), in a library, or doing more traditional historical research and teaching in a university. Traditional historical research involves looking at primary texts (books, letters, maps, drawings), knowing a second language in order to read foreign language materials without an English translation, and interpreting evidence to draw conclusions about how cultural and social interactions affected the development of science or technology. Historians with a scholarly focus typically teach at a college at the undergraduate or graduate level and have a Ph.D. in the History of Science. High school science or history teachers might have taken classes in the history of science. Classes such as history, English, biology or other sciences, foreign languages, and computers are recommended.

Turning Corn into Milk: Alchemy or Biochemistry?

PROLOGUE A cow stands over the grain bin contentedly chewing on corn. In a few hours, the dairy farmer will come in to milk her. What is the relationship, if any, between the corn that provides the cow with nutrients and energy and the milk that she produces both for feeding her offspring and for the farmer to sell?

In this learning experience, you will use the concepts from Learning Experience 5 to examine how organisms use food resources to synthesize new materials for their own use; you will identify the chemical relationship between the food that a cow uses (corn) and one of the products a cow synthesizes (milk) and determine how one can be turned into the other.

Corn and Milk: So Different Yet So Similar

INTRODUCTION How can a cow use a food resource like corn to produce milk—a very different substance? In this investigation you will analyze the biomolecular composition of corn and milk, using indicators to determine whether corn can provide any of the components found in milk.

The following investigation uses the indicators Biuret reagent, Benedict's solution, iodine, and Sudan III to test for proteins, sugar, starch, and lipids. Look at the investigation you conducted in Learning Experience 4—Feeding Frenzy, and review the use of indicators and the general procedure for conducting an indicator experiment.

MATERIALS NEEDED

For each group of four students:

- nutrient label from a milk carton
- 4 safety goggles
- 4 pairs of disposable gloves (optional)
- 1 eyedropper or 10 disposable pipettes
- heat source
- 2 beakers (250-mL)
- 12 test tubes
- 1 test tube clamp
- 1 test tube rack
- 1 glass stirring rod
- 3 teaspoons canned whole-kernel corn
- 10 mL fresh, whole, pasteurized milk
- 1 small bottle Biuret reagent
- 1 small bottle Benedict's solution
- 1 small bottle iodine
- 1 small bottle Sudan III or IV (or alternative test as described in Learning Experience 4)
- 1 fork or spoon (to mash corn)
- 1 wax marking pencil

For the class:

- 1 set positive test control solutions
- 1 teaspoon
- boiling water bath
- blender (optional)

SAFETY NOTE: *Always wear safety goggles when conducting experiments.*

PROCEDURE

1. STOP & THINK Identify the question being asked and make a prediction about the outcome of this investigation. Write your response in your notebook.

2. STOP & THINK Read through the entire procedure and draw your experimental setup in your notebook. See the illustration of the setup in the Procedure section of Learning Experience 4 for an example.

3. STOP & THINK Why is it important to analyze milk even when you have access to information about milk from the nutrition label? Record your reasons.

4. Take 3 spoonfuls of corn with its juice. Place in a 250-mL beaker (or clean plastic container) and mash up the kernels with the back of a fork (or you may use a blender).

5. Pour approximately 10 mL of fresh milk into a container.
6. Label a set of four test tubes: Water +, then label each with one of the following, "sugar", "starch", "protein", or "lipid." These are your negative controls. Place the test tubes in the test tube rack. There will be a class set of positive controls.
7. Label a second set of four test tubes: Milk +, then label each with one of the following: "sugar", "starch", "protein", or "lipid."
8. Label four more test tubes: Corn +, then label each with one of the following: "sugar", "starch", "protein", or "lipid." Place the test tubes in the test tube rack. These last two sets are your experimental unknowns.
9. Using a clean pipette or eyedropper, add 30 drops of water to each of the test tubes labeled "water."
10. Using a clean pipette or eyedropper, withdraw the corn mush from the beaker and place 30 drops of the mush into each of the test tubes labeled "Corn."

    If you are using an eyedropper, be sure to rinse the eyedropper well between each step. To rinse: draw distilled water from one beaker into the dropper, then squirt the water into a second beaker. Repeat 2–3 times for thorough rinsing.)
11. Using a clean pipette or eyedropper, place 30 drops of milk in each test tube labeled "Milk." Wash the eyedropper.
12. Using a clean and separate pipette for each indicator or a clean eyedropper, add 15 drops of the appropriate indicator to the appropriate test tubes. You should have three test tubes for each indicator: two samples (corn and milk) and one negative control. Heat in boiling water those test tubes that contain Benedict's solution.
13. Examine all your test tubes for any color changes and record the results. You may need to hold the Biuret test tubes against white paper to see the color changes more clearly.
14. Record your results. Discuss them with your group.
15. Dispose of the contents of the test tubes and wash your glassware.
16. Write a laboratory report that includes the following:
    - the question(s) being asked;
    - your predictions about the outcome;
    - the experimental design and how the investigation was set up, including a rationale for testing milk again;
    - the data or observations you made (in a chart or table);
    - your analysis (your responses to the Analysis questions);
    - your conclusion (your answer to the question being asked).

CAUTION: *The hot plate and boiling water can cause burns. Biuret is caustic. If it comes in contact with skin, rinse thoroughly with water. If you should get any in your eyes, irrigate them immediately and inform your teacher.*

ANALYSIS

Write responses to the following questions in the analysis portion of your laboratory report in your notebooks:

1. What biomolecules are present in milk and corn? What biomolecules make up a cow? How do you know this?
2. What chemical elements (atoms) make up these biomolecules?
3. What chemical elements are needed to make a protein? Might it be possible to make proteins if you don't take in proteins? How might this be possible?
4. How can two different substances—corn and milk—be composed of many or possibly all (depending on your results) of the same biomolecules?
5. Based on the results of your experiment, create a diagram, in words or drawings, illustrating the pathways by which a cow uses corn to make milk.

HAVEN'T I SEEN THAT CARBON SOMEWHERE BEFORE?

Julie Lewis walks on a dream come true. Since she was a teenager, inspired by the rallies of the first Earth Day in 1970, she yearned to turn waste into something worthwhile. Now the 38-year-old is vice president of a company she founded called Deja. She calls her recycled invention the Deja Shoe.

Its cotton-canvas fabric is rewoven from textile scraps. The foam padding was designed to cushion chairs. Factory-reject coffee filters and file folders go into the insole. Add recycled grocery bags, tire rubber, and plastic trimmings left over from the manufacture of disposable diapers.

The shoes look handsome, durable, and ready for the outdoors... Her Portland, Oregon firm ships 100,000 pairs annually [to] stores across the country. And when they wear out? Send them back to Deja to be recycled.

—*An excerpt from Noel Grove. "Recycling."* National Geographic *July, 1994, pp. 92–115*

Julie Lewis's company takes trash and turns it into shoes; cows take corn and use it to produce milk. Similarly, you eat pizza or salad and convert them into proteins, lipids, carbohydrates, nucleic acids, and energy that you need in order to continue growing and maintaining your body. Long before the first Earth Day, organisms were recycling materials.

USE IT AGAIN, SAM

How do living organisms recycle the resources they take in, and transform them into new materials they can use? How can a cow use a food resource like corn to produce milk—a very different substance?

As you have been finding out, perhaps corn and milk are not totally different substances, at least in terms of the biomolecules and chemical elements of which they are composed. The big differences are in how those elements are put together. For example, all biomolecules are composed of similar elements. But the biomolecules differ because each of their elements are put together into different subunit structures.

Because these subunits can also be varied, many possible kinds of carbohydrates, proteins, nucleic acids, and fats can be made. Single subunits can be different, and groups of subunits can be put together in lots of different ways. These sometimes slight, sometimes more obvious differences in the way things are put together can produce materials with very different structures, chemical properties, and functions. The carbohydrate in milk, for instance, consists primarily of the disaccharide sugar lactose, whereas the primary carbohydrate in corn is the polysaccharide starch.

Grass, another food source for cows, has another polysaccharide, cellulose, as its major carbohydrate. Starch and cellulose are both composed of a long chain of linked glucose molecules; these two carbohydrates differ only in terms of how those glucose subunits are linked to one another. Lactose is a simpler carbohydrate composed of a glucose molecule and a galactose molecule. Glucose and galactose are composed of exactly the same number of carbons (6), hydrogens (12), and oxygens (6), ($C_6H_{12}O_6$) but they differ in how these atoms are arranged (see Figure 6.1). This small difference in arrangement produces two sugars that are chemically different in nature. When these two sugars are joined by a chemical bond, yet a third and different sugar, lactose, is formed.

Imagine that you are an architect who has been hired to design a building constructed only of wood, ceramic, and glass. Using only these

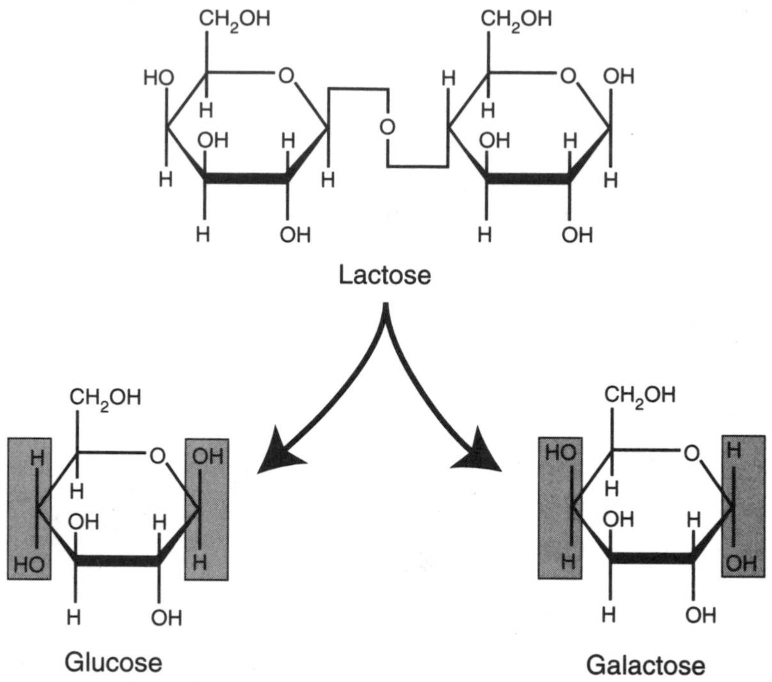

Figure 6.1
Lactose is a combination of glucose and galactose

three materials, you have a tremendous variety of possibilities for how the final building might look. Just as carbon, hydrogen, and oxygen can be put together differently to make substances very different in nature from one another, you could design a variety of different buildings by changing the way building blocks are put together.

Although all living organisms are made up of *organic* (carbon-containing) materials that have the same six elements, life on Earth is diverse. This is due to the highly varied possibilities for design in how organic compounds can be put together. The similarity in the elements that living things are made of and the variations in the structure of biomolecules suggest a start to responding to our question of "How do living organisms recycle the resources they take in and transform them into materials they can use?" If biomolecules differ in terms of how their elements are arranged, then perhaps one could be made into the other by reshuffling the chemical elements. If a cow could break down corn into smaller subunits, it could then rearrange them and build something new—similar to Julie Lewis taking scrap pieces of textiles and weaving them together into a whole new shoe.

BREAKING IT DOWN, BUILDING IT UP

These chemical reactions of breakdown (*catabolism*) and synthesis (*anabolism*) are fundamental to how living organisms sustain life. One definition of a chemical reaction can be the transformation of molecules into other kinds of molecules. Anabolic reactions involve the building of biomolecules from other molecules. You have already seen an example of this in photosynthesis, in which carbon dioxide and water are transformed into sugar. The transformation of corn into milk actually involves many steps in which the starch and other biomolecules are broken down (catabolized) and the components reassembled into the sugars and other biomolecules of milk. Figure 6.2 illustrates the relationship of anabolic reactions to catabolic reactions.

How do biomolecules get "broken down"? One of the major requirements is energy. If you have ever burned toast, you know that it turns black. The heat provided the energy that caused some of the bread to break down to carbon. This can also be seen when glucose is "burned." When heat energy is added in the presence of oxygen, the glucose molecule breaks and a great deal of energy in the form of heat is released. Chemically, electrons are transferred from the hydrogen atoms in the glucose molecule to oxygen, resulting in the formation of water. When a molecule loses electrons, it is said to be *oxidized*. The chemical reaction for the burning of glucose can be written as the following equation:

$$C_6H_{12}O_6 + 6\ O_2 + \text{energy} \longrightarrow 6CO_2 + 6H_2O + \text{energy}$$

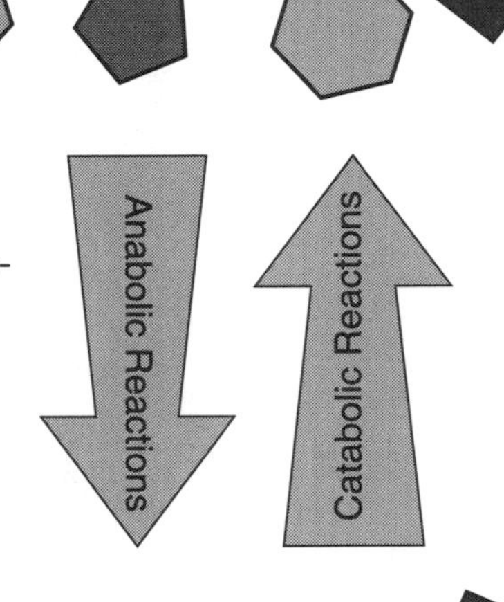

Figure 6.2
The cycle of catabolism and anabolism

In living organisms, a very similar reaction occurs but in a much slower, controlled, series of small steps. The slower oxidation of glucose has several advantages for the organism; it prevents the organism from burning up; it enables the energy in the bonds of the glucose molecule to be transferred to other molecules for use by the organism rather than being released as heat; and it allows the products of the glucose molecule breakdown to be used by the organism to synthesize other biomolecules. In the next learning experience, you will be examining these energy transfers and the importance of oxygen in them.

PUTTING IT ALL TOGETHER

Metabolism is the web of simultaneous and interrelated chemical reactions taking place at any given second of life. Within this web, complex chains of biomolecules are woven from simpler units or are dismantled piece by piece. Growth, movement, repair, and other life-sustaining activities depend on these collective chemical reactions.

Metabolic reactions are organized into pathways. Catabolic reactions are interconnected to anabolic pathways; the product or *intermediate* of a catabolic pathway may be the starting material for an anabolic pathway. The anabolic pathways consist of a number of individual steps through which materials are progressively rearranged and built. These sets of reactions are termed pathways because they have a starting material (for example, starch), an end product (protein, nucleic acid, lipid, carbohydrate), and a series of reactions in between (steps along the path). At each step of a metabolic pathway, the starting material is changed a little more (see Figure 6.3). Sometimes a new pathway can begin in the middle of an ongoing pathway.

In your investigation you determined that milk and corn contained the same kind of biomolecules. The cow ingests corn; the corn is broken

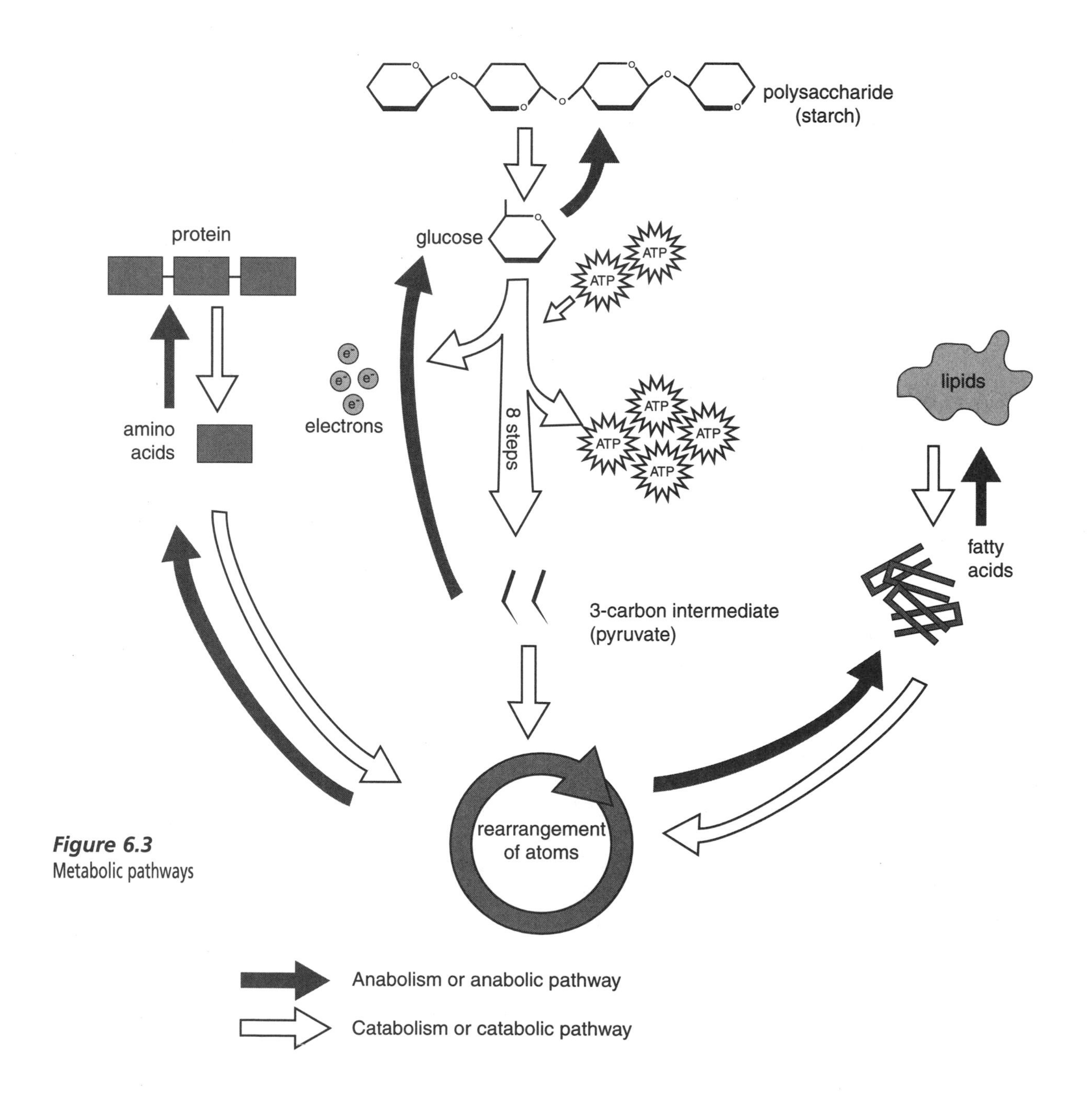

Figure 6.3
Metabolic pathways

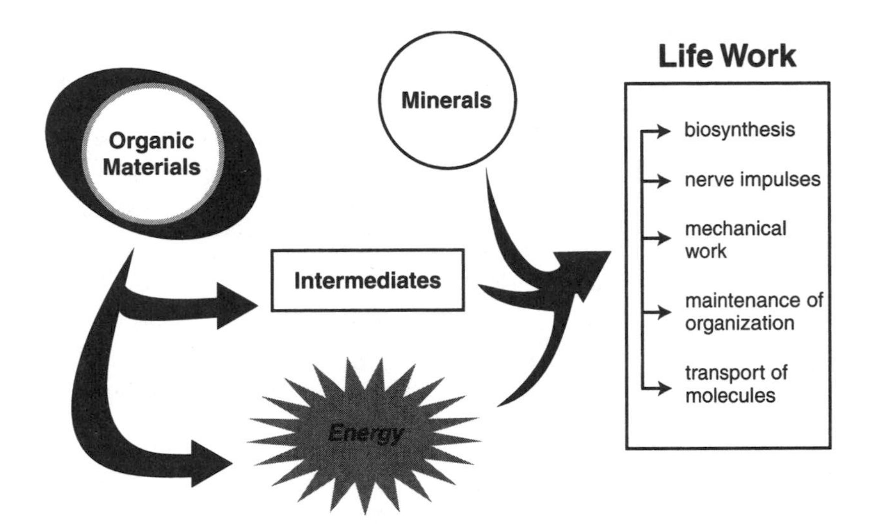

Figure 6.4
Overview of metabolism

down further and further, first by the digestive processes into its biomolecular components and then by the catabolic pathways, into intermediates. This process releases energy which is then captured in a chemical form that the cow can use. These intermediates are then used as building materials to synthesize new biomolecules using the energy from catabolism. These materials are also used for the other life processes the cow must carry out. Figure 6.4 represents these transformations.

ANALYSIS

Create a metabolic pathway flow chart that begins with corn and includes concepts from the reading. Use the illustrations to guide you in your diagram and as an aid in annotating what is happening in each step.

EXTENDING IDEAS

- The concept of recycling is a popular one today, whether it is making shoes out of garbage or finding uses for the packaging materials that enclose many items that we purchase. Blue or green bins in many cities contain various items for recycling, including glass, aluminum, polystyrene foam, and plastic. Write or draw an analogy of the recycling of food packaging material as it compares to the breakdown and synthesis of food; describe what happens to the package during the recycling process and how this compares to the recycling of biomolecules.

ON THE JOB

TEACHER Do you like sharing what you know about flowers, plants, or animals? Teachers combine their knowledge of the subject with the ability to make it interesting and to communicate the knowledge clearly. Teachers at all grade levels plan what topics to cover and what activities or labs to use to illustrate the topic, prepare activities or labs (this includes collecting all the materials and setting up the activity), work with students as they do the activities, and write and grade quizzes and tests. A teacher who works in an elementary school might teach one class all subjects, including science, or could just teach science and work with all the classes in the school. Elementary school science consists of lots of activities and of encouraging students to ask a lot of questions and wonder about the world. A middle school science teacher teaches general science courses which focus on science as a way of thinking and continues to build on the concepts students explored in elementary school. High school science teachers most often teach in one particular subject area (biology, chemistry, or physics), although many are versatile and can teach in several subject areas. High school science explores science concepts in depth and also looks at the science within social, historical, ethical, and political contexts. Not all science teachers work in a school. Some work in museums, zoos, aquariums or science centers teaching the general public and leading classes on school field trips. A minimum of a four-year college degree is required with classes in your subject area and the completion of a teacher preparation program. The teacher preparation program includes courses on child development, on teaching methods and student teaching. Requirements for teacher certification vary from state to state. Some teachers have a master's degree in education or science education. Classes in subjects such as biology, chemistry, mathematics, English, and psychology are useful.

A Breath of Fresh Air

PROLOGUE Take a deep breath. What did you just inhale? Air, which is made up of gases, is an essential resource for most living organisms. In searching for life on other planets, biologists use changes in atmospheric gases as an indicator of a life presence. In the investigation in Learning Experience 3, the plant grown in the absence of carbon dioxide was unable to carry out photosynthesis, one of its life processes. The plant could not transform carbon dioxide and water into sugar. How do other organisms use air? How is air involved in the metabolic processes that you explored in the last learning experience? In this learning experience, you will examine which components of air are required by living organisms, how and why the composition of the air you inhale is different from that of the air you exhale, and the role of air in the metabolic processes. What part of the air are nonphotosynthesizing organisms using and what changes are occurring as organisms "breathe" or respire?

What Goes In...

INTRODUCTION What happens when we take air into our bodies? Does it change composition or remain the same? The composition of air can be determined using indicators. In the following activity, you will analyze data in which the gases of inhaled and exhaled air are compared.

TASK

Using the data from the table on the next page (Table 7.1), write responses to the following questions:

1. How do you think this data was obtained?
2. What does the data tell you about which component of air we use?

3. Why do you think the carbon dioxide content increased?
4. What does the nitrogen data tell you?
5. How might you design an experiment to answer the question "What happens to air in the presence of living organisms"?

Table 7.1
Analysis of air inhaled and exhaled by humans

GAS	INHALED	EXHALED
Oxygen	19.70%	15.70%
Nitrogen	74.10%	74.50%
Carbon Dioxide	0.04%	3.60%

...Must Come Out

INTRODUCTION Can we measure changes in the composition of air as living organisms use it? What do living organisms use from the air they take in? In the following investigation you will determine which gases organisms use. You will use chemical indicators to observe changes in the air that occur in the presence of living things and explore why these components of air are needed.

MATERIALS NEEDED

For each group of four students:

- 4 safety goggles
- 9 large test tubes
- 3 small test tubes that fit inside the larger tubes
- 3 cork stoppers to fit the large tubes
- 1 test tube rack
- 2 wrapped straws
- paper towels
- 1 small bottle phenol red solution
- 1 small bottle limewater
- 1 small bottle carbonated water
- 1 small bottle vinegar
- 1 wax marking pencil
- 1 eyedropper
- 2 beakers (250-mL)
- access to distilled water

- One of the following pairs:
  - yeast-sugar solution
  - boiled yeast-sugar solution

  OR

  - 1 small live insect
  - 1 small dead insect

▶ PROCEDURE

SAFETY NOTE: *Always wear safety goggles when conducting experiments.*

1. **STOP & THINK** Identify the question being asked in the experiment and create a hypothesis. Predict the results of the experiment ("If.....then....").

2. **STOP & THINK** Read through the procedure and determine how you will record your data.

   Before beginning, record the following:
   - the variable in the experiment
   - the positive and negative controls (HINT: As in many scientific investigations, there are many more controls than experimental samples. In this experiment, there are 7 controls. To identify each control, copy the test tube setup diagrams into your notebook and indicate which ones are controls for which indicators: one control that is negative for acids, three that are positive for one indicator, one that is negative for one indicator and positive for the other indicator, and two that are positive for both indicators.)

3. Use the wax marking pencil to label nine large test tubes 1, 2, 3, 4, 5, 6, 7, 8, and 9. Place them in a test tube rack.

4. With an eyedropper add 10 drops of phenol red solution to each of the nine test tubes. Be careful not to touch the side of the test tube with the eyedropper. Hold the dropper over the mouth of the test tube and let drops fall freely. Phenol red is an indicator that changes color in the presence of acids.

5. Rinse the eyedropper. To rinse: draw distilled water from one beaker into the dropper, then squirt the water into a second beaker. Repeat 2–3 times for thorough rinsing.

6. Pour limewater into tubes labeled 1, 2, 3, 7, 8, and 9 until each is about one quarter full. Limewater indicates the presence of carbonic acid (a weak acid produced by carbon dioxide and water).

   CAUTION: *Limewater is an irritant. Avoid touching it. If it comes in contact with skin, rinse immediately with water.*

   Unlike the investigations in Learning Experiences 4 and 6 in which one indicator was used at a time, in this investigation, two indicators are being used in the same test tube; phenol red which detects all acidic solutions (such as vinegar), and limewater which detects specifically carbonic acid (formed when carbon dioxide is dissolved in water).

7. Label the small test tubes 1, 2, and 3. In the small tubes place the following materials:

   Test tube 1: 10 drops of water

   Test tube 2: a small, rolled piece of paper towel soaked in yeast-sugar solution

Test tube 3: a small, rolled piece of paper towel soaked in boiled yeast-sugar solution

OR

Test tube 2: small live insect

Test tube 3: small dead insect

8. Slide the small test tubes gently into their matching numbered large test tubes, (see Figure 7.2) then place corks in all three large test tubes. Observe and record initial observations in your notebook.

Figure 7.2
How to set up test tubes 1–3

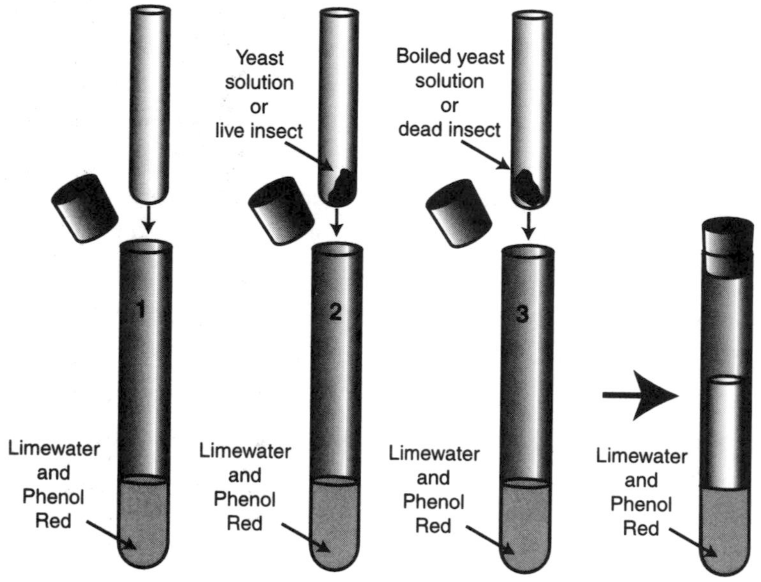

9. Add the following directly to the large test tubes labeled 4, 5, and 6 (see Figure 7.3):

Test tube 4: 5 drops of vinegar (3–5% acetic acid)

Test tube 5: 5 drops of carbonated water (carbonated water contains dissolved carbon dioxide)

Test tube 6: your breath, blown through a straw for 30 seconds into the phenol red solution

Be sure to rinse your eyedropper between **each** step.

CAUTION: *Do not suck any liquid back up the straw.*

10. Add the following to test tubes labeled 7, 8, and 9 (see Figure 7.4):

Test tube 7: 20 drops of vinegar

Test tube 8: 10 drops of carbonated water

Test tube 9: your breath, blown through a straw for about 30 seconds, into the limewater and phenol red solution

Be sure to rinse your eyedropper between **each** step.

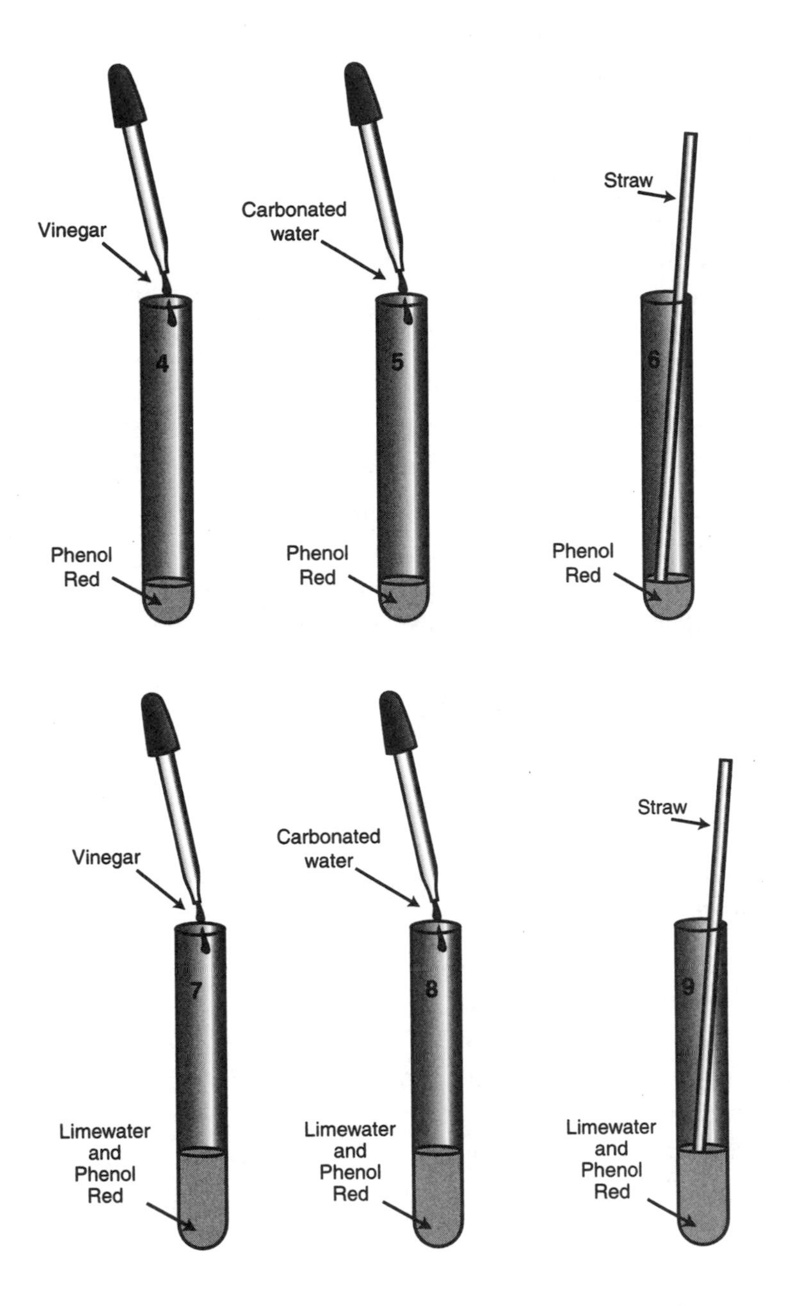

Figure 7.3
Setup for test tubes 4–6.

Figure 7.4
Setup for test tubes 7-9

11. Create a table in your notebook in which you can record your observations of each test tube. Record the results from your indicators (color and other changes) from all nine tubes. Compare your observations in tubes 2 and 3 with the other tubes. Compare your observations in tubes 2 and 3 with your first observations made in step 7.

ANALYSIS

1. Using your data:
   - determine whether your hypothesis is correct. Explain why you think your hypothesis did or did not agree with your results.
   - identify one characteristic of living organisms based on your experimental samples in test tubes 2 and 3. Explain your conclusion.
2. Using your data and the data in Table 7.1:
   - predict what kind of change might be observed in the oxygen content of the air in test tube 2; explain your prediction;
   - describe or diagram how you think these changes in the composition of air might occur in the presence of living organisms. Indicate what additional information you might need in order to explain what has caused the change in the composition of the air.

ENERGY FOR ALL

What would happen if we left an insect in a stoppered bottle for a length of time, even in the presence of food and water? Why does it need oxygen? Why does a lack of oxygen quickly lead to death? Clearly, oxygen is essential to life.

FROM SUN TO LIFE

Much of your exploration so far has focused on the need of living organisms for energy and building blocks to sustain life. But where does this energy come from? As you learned in Learning Experience 3, the origin of energy on Earth comes from the sun. Plants transform solar energy into chemical energy (sugar). Plants then use this sugar to form carbohydrates, proteins, lipids, and nucleic acids through the processes of anabolism and catabolism.

During metabolic processes, the energy from the sun flows, in the form of chemical energy, from molecule to molecule and is used to take molecules apart and build new molecules. During catabolism larger molecules are broken down or oxidized. This reaction requires the input of some energy, but much more energy is released from the chemical energy stored in the chemical bonds in the molecules. During anabolism energy is stored in chemical bonds where it remains until the molecule is catabolized in another reaction. Thus, metabolism is two interconnected processes: that of breaking down molecules to make building blocks which produce energy, and that of building up molecules that store the energy needed to carry out functions in the cell.

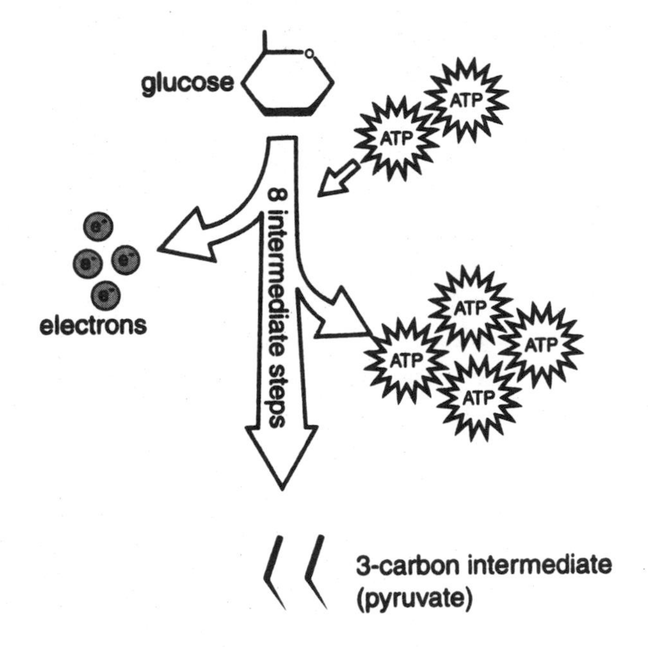

Figure 7.5

Energy transfer during glycolysis. In order to break the bonds in glucose, the energy of two ATP molecules is required. During the sequential steps between the breakdown of glucose and the formation of two pyruvates, four molecules of ATP are formed.

IT TAKES ENERGY TO MAKE ENERGY

As described in Learning Experience 6, the breakdown or oxidation of glucose is a multi-stepped process in living organisms. The bonds in glucose are broken, not in a single step as in the burning of glucose, but in nine steps. The bonds in the glucose molecule are broken using chemical energy. The products of this breakdown are rearranged in eight steps to form eight intermediate molecules. The end products of this breakdown of glucose (termed *glycolysis,* meaning "glucose breaking"), is the generation of two new molecules that contain three carbons each (*pyruvate*) and energy that is released from the chemical bonds of glucose. This energy is transferred to a new molecule, *adenosine triphosphate* or *ATP*.

ATP is a very important molecule in living organisms. It is widely used as an energy "holding tank" in most living organisms. This molecule holds the energy released during the catabolism of glucose until it is needed by other processes in the cell such as anabolism and catabolism. Figure 7.5 is an overview of glycolysis showing the starting material glucose and the eight intermediate products before the 3-carbon molecules are formed. In addition, it shows that the reaction requires an input of two ATP molecules, but that four molecules of ATP are formed by the process. That means that there is a net gain of two ATP molecules available for other work of the organism. Although a net gain of two ATPs was made, in the bigger picture of the energy needs of a living organism it is not nearly enough. In addition, the energy transferred from glucose at this stage to ATP represents only about two percent of the energy in the original glucose molecule. How does an organism tap into the rest of that energy?

The end products of glycolysis, those 3-carbon molecules called pyruvate, enter into yet another pathway. In this pathway, these pyruvate molecules are broken down, rearranged, and rejoined in a series of steps to generate new molecules that form the starting material for other anabolic pathways. During the course of this pathway electrons are transferred (remember, chemical reactions are actually the transfer of electrons) from molecule to molecule.

This transfer of electrons also involves the transfer of energy that is contained in the electrons. It is the energy in the electrons that is ultimately transferred to ATP. Carbon dioxide is released at this point as a waste product. From the beginning of the breakdown of glucose, a net gain of four ATP molecules has been made (see Figure

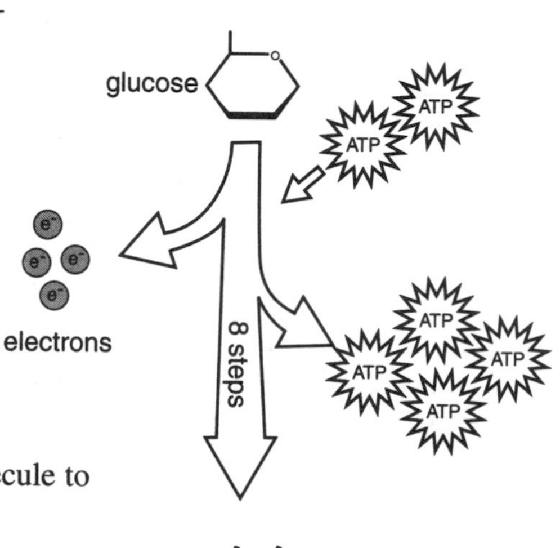

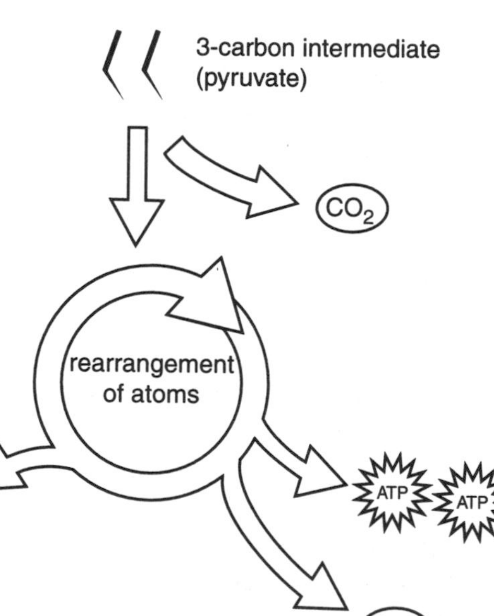

Figure 7.6

Two more ATPs. Pyruvate formed by glycolysis enters another metabolic pathway in which it is transformed through a series of steps that result in the net gain of two molecules of ATP and the release of two molecules of carbon dioxide as byproducts.

7.6). Still not enough! However, during this pathway many high-energy electrons have changed "hands" (actually molecules) and there is still a lot of energy to be had. How can the organism get more of this energy converted into ATP?

LET'S GET OXIDIZED!

Oxygen is the organism's key to obtaining, finally, the solar energy that was trapped in glucose. During the rearrangement in the two pathways shown in Figure 7.6, high-energy-containing electrons were transferred to molecules whose main purpose is to pass them on, through a complex transport system. The energy that is present in these electrons when they enter this transport system is gradually transferred, in a series of steps, to form ATP. Oxygen, which organisms have taken from the air, is the final acceptor of the electrons. The net gain of ATPs from this transfer of electrons is 32. Much better!

At the end of these reactions, the electron has lost its high energy by passing it on to the molecule of ATP. Finally, we see a role for oxygen. In any system the role of waste remover is essential, and oxygen serves that essential role for living organisms by removing the spent electrons that are byproducts or waste products of metabolic processes. The final product of this reaction is water (see Figure 7.7). Thus, two waste products are generated by the catabolic reactions just examined, carbon dioxide and water. The products gained by these reactions are building materials for the anabolic reactions and energy in the form of 36 molecules of ATP. This transport of electrons to oxygen and the generation of ATP along the way is referred to as *respiration.*

The rapid death of most living organisms when deprived of oxygen, then, is a direct result of the organism's need for an electron acceptor. If oxygen is not present, the flow of electrons is stopped and that large quantity of ATP can not be made. Organisms need a constant supply of energy to continue to maintain the characteristics of life. In the absence of oxygen, the supply of energy is rapidly depleted and the organism can no longer carry out its energy-requiring life functions. Figure 7.8 presents an overview of metabolism showing the pathway's form, energy transfer, anabolism, and catabolism.

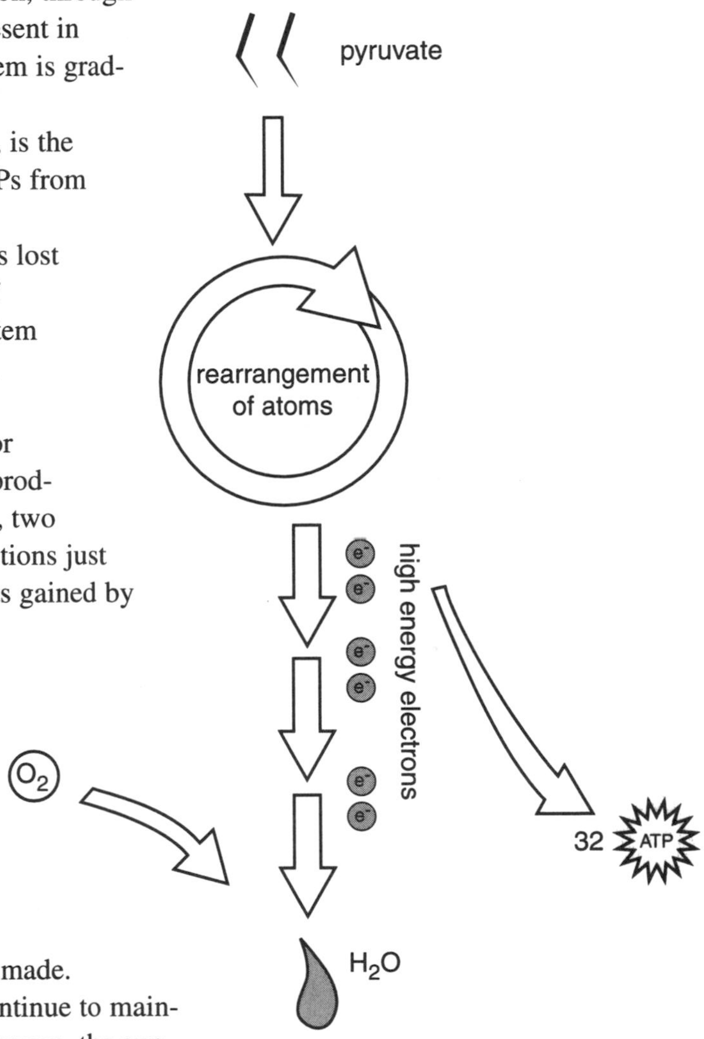

Figure 7.7

High energy electrons are passed through a series of reactions in which oxygen serves as the final electron acceptor. Then two more ATPs are formed and water is formed as a byproduct.

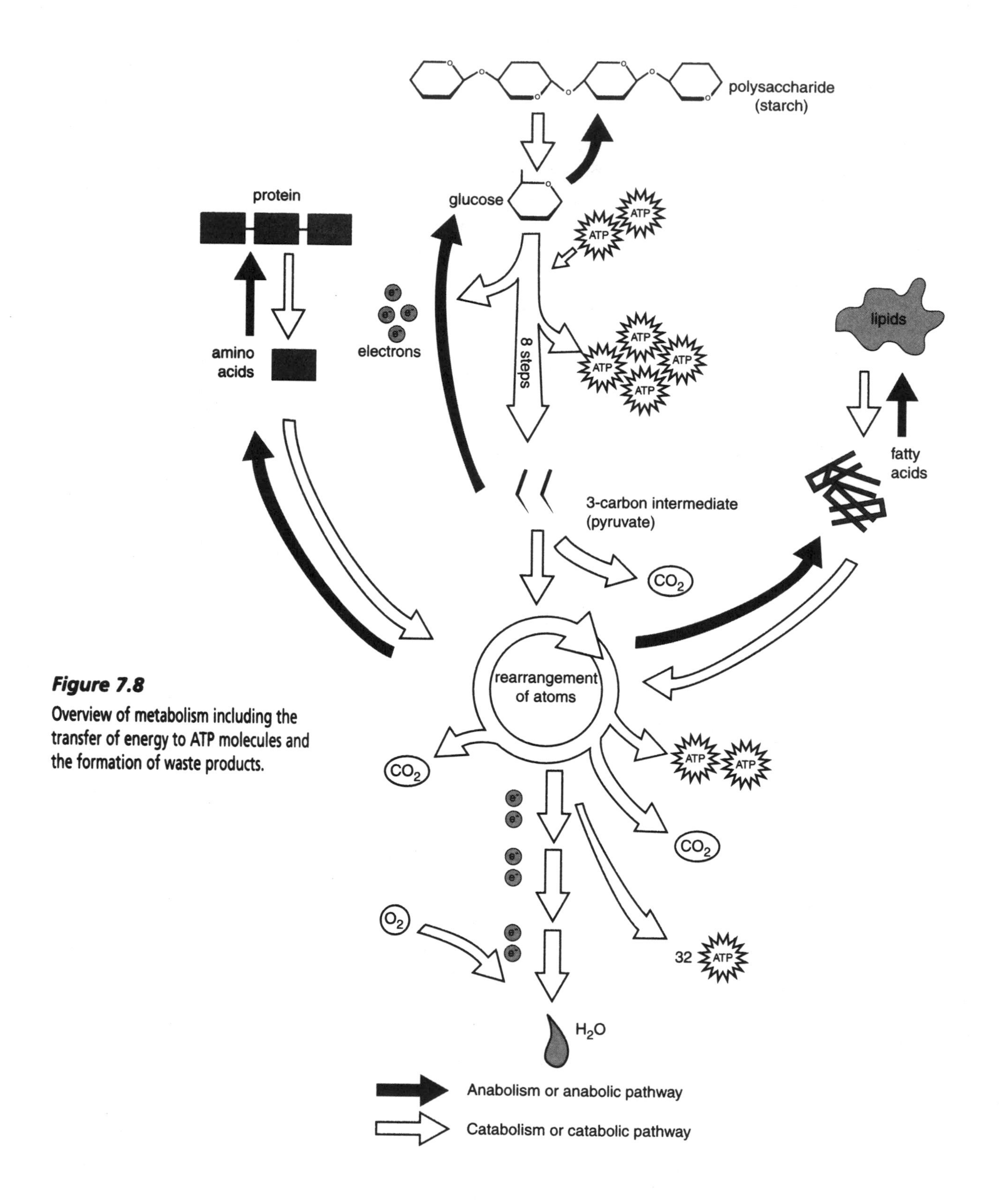

Figure 7.8

Overview of metabolism including the transfer of energy to ATP molecules and the formation of waste products.

ANALYSIS

1. In Learning Experience 6 the chemical reaction for the burning of glucose was described as:

$$C_6H_{12}O_6 + 6O_2 + \text{energy} \longrightarrow 6CO_2 + 6H_2O + \text{energy}$$

   In words or drawings explain how this reaction occurs metabolically. Include:
   - where the glucose came from;
   - where the oxygen came from;
   - the fate of the glucose molecule as it is catabolized;
   - the fate of the oxygen molecule;
   - any products which are formed as a result of intermediate reactions;
   - where the energy for the catabolism is obtained;
   - an indication of the net energy gain;
   - a comparison (differences and similarities) between burning glucose and catabolizing glucose.
2. Do you think that photosynthetic organisms carry out this process? Why or why not? Where would the glucose come from?
3. In 1772, Joseph Priestley observed that "good air" supported the burning of a candle flame. He also observed that "good air" supported the breathing of an animal. If a mouse was placed under a jar it would live for a while and then die. If a plant was placed under the jar with a mouse, the animal lived longer than it would have without the plant. Explain the science in each of these results.
4. The poison cyanide inhibits oxygen consumption in living organisms. It acts by blocking the final transfer of electrons to oxygen which effectively stops the production of ATP. Describe or draw the effect that this poison has on living organisms and why.
5. Describe how the process of respiration might cause the changes in the air that you observed in "What Goes In..." and "...Must Come Out."

EXTENDING IDEAS

- Certain organisms can live and maintain life in the absence of oxygen by generating energy by a metabolic process called fermentation. Yeast cells carry out a fermentation process whose byproducts are used in several ways. Describe the chemical reactions involved in the process of fermentation including the starting resources yeast use, how yeast obtain energy and building blocks during this process, and how the byproducts of these reactions have been used by humans. Identify other organisms that provide products useful to humans through the process of fermentation.

ON THE JOB

SCIENTIFIC ILLUSTRATOR Are you intrigued by science, medicine, and technology, but find your true passion to be drawing and illustration? Scientific illustrators combine their talent and technical ability in illustration with their scientific background to produce the charts, graphs, illustrations, or cover art that accompany scientific articles. Scientific illustrators might be employed by a company or act as freelance artists working on several projects with scientific laboratories in universities or in businesses, magazines (both professional journals and for the general public), newspapers, or publishing companies. The illustrator's grasp of the subject matter is used to talk with the doctors or scientists who are asking for the art work. Illustrations are created using computer graphics, as well as by hand drawing. There are some types of illustrations (such as medical illustrations) that are done better by hand than using the computer because the artist is better able to portray the level of detail necessary. Scientific illustration is a specialty of illustration and drawing because of its focus. Medical illustration is a specialty of scientific illustration. Medical illustrators enter special degree programs and in addition to computer graphics and drawing courses complete the first year of medical school which includes anatomy classes and cadaver dissections. Scientific illustrators usually have at least a college degree. Classes such as biology, chemistry, physics and other sciences, drawing, illustration, computer graphics, business courses, and English are useful.

POLLUTION CONTROL TECHNICIAN Are you concerned with the quality of the air you breathe or the water you drink? Pollution control technicians conduct tests and field investigations to determine ways to both monitor and control the contamination of the air, of fresh water or soil. Two major specialties of technicians are air pollution control and water pollution control, although some technicians may specialize in noise, light, or soil pollution. An air control technician might collect and test air samples, record data on atmospheric conditions, study specific chemical or waste products that contribute to pollution such as automobile engines, replenish chemicals used in tests, replace parts, calibrate instruments, record results, or read and interpret maps, charts or diagrams. To test the quality of outdoor air, sampling devices that detect the amount of solid particles and of certain gases are typically installed on rooftops, stationary trailers or on mobile trailers (which are like mini-labs). Technicians usually have an understanding of the environment (the air, water, soil, etc.) in which the data is collected in order to detect the presence of gases or particles that are not normally present and discriminate between the importance of pollution factors. Technicians usually have either a two year college degree in the physical sciences or two year post high school training in pollution control technology. Since this is a relatively new field, some positions with on the job training are available to those with a high school diploma and appropriate experience. Classes such as mathematics, chemistry, physics, biology (conservation or ecology), computer courses, and English are useful.

REAP THE SPOILS

PROLOGUE The very basics of sustaining life rely on a constant exchange of energy and chemical compounds among living organisms. There is a balance to this exchange; a product or waste of one organism may be a necessary resource for another. Because this flow of energy and matter links all living organisms, any change to any connection can alter the balance of the entire system.

Over the course of the next two weeks, you are going to observe and analyze these kinds of connections within a model environment, milk. How is milk an "environment"? As you have seen in past learning experiences, foods contain nutrients that organisms use to sustain life. Milk has many nutrients, some of which you have identified. If milk is left open to the air, even for a short time, it can become the environment for many different kinds of microorganisms that are floating around in the air, searching for new environments with nutrients they can use as resources. As the organisms grow and live, dramatic changes can occur within this environment.

HOME SWEET MILK

INTRODUCTION You will begin the investigation by setting up your model environment. Once you set up the environment, you need to design your investigation to include careful observations over time and a method to record detailed and accurate data. The data should include: kinds of organisms living in milk, how these populations alter the environment in which they live, and how they affect one another. This investigation will continue over the course of the next two weeks. Every 2–3 days you will spend a class session sampling and testing your milk environment and talking with your group members about what you observe.

MATERIALS NEEDED

For each group of four students:

- 4 safety goggles
- 125 mL whole pasteurized milk
- 1 beaker (250-mL) or clean plastic margarine or other container
- litmus or pH paper or access to pH probe
- 1 thermometer
- 4 nutrient agar plates
- cellophane tape
- 4 sterile cotton swabs or 1 inoculating loop
- 8 microscope slides and coverslips
- 1 eyedropper
- distilled water
- 1 wax marking pencil
- plastic wrap
- 1 dissecting needle or fork
- 1 flame source (candle or Bunsen burner)
- 3 empty petri dishes or plastic containers for holding water
- crystal violet stain
- paper towels
- 1 forceps or tweezers
- water

For the class:

- compound microscopes
- 1 pH probe (optional)
- 1 32-37°C incubator (optional-can be made from cardboard box)
- 1 pot holder or oven mitt
- 1 microwave oven or other heat source

PROCEDURE

1. 

Determine the question being asked, write a hypothesis, and predict the outcome of this investigation.

2. 

What kinds of tests and techniques would you need to carry out in order to determine what happens in the milk environment over time? Use your understanding of metabolism from last few learning experiences to help determine some of the tests and techniques you might want to include. In the following reading there are descriptions of techniques for:
   - observing populations
   - measuring pH

Use these techniques, in addition to others you might include in your experimental design. Don't forget to use your senses as indicators of change in your experimental design.

3. STOP & THINK Develop ways of illustrating your observations and data-collecting system that are clear, accurate, and consistent. Develop ways of keeping track of questions that arise in the course of the investigation and ways of recording any unexpected turn of events that may affect your outcome and results. (For example, knocking over your container of milk.) Your data and illustrations and the conclusions you draw from the data and illustrations will be needed to complete your laboratory report and for a class presentation.
4. Write a procedure that includes all the necessary information from the procedural steps 1-3 above. (Every individual in the group will need to keep his or her own record of the data and his or her own illustrations.)
5. Once your teacher has approved the experimental design, obtain an empty 250-mL beaker (or margarine container) and label it with your group's name. Pour 125 mL of whole milk into the beaker.
6. Take any samples, do any tests, and record any measurements or observations your group has decided upon. Record your data in your notebook.
7. Find a storage location in the classroom where you can leave your milk sample undisturbed for several days. Check with your teacher on this location. Take careful note of the characteristics of the area where your sample will be incubating: whether it is sunny, cold, drafty, near a window, over a radiator, etc.
8. Place your milk sample, uncovered, and the covered agar plate (inverted, agar side up) in the designated storage location.

SAFETY NOTE: *Always wear safety goggles when conducting experiments.*

TECHNIQUES

Before beginning the experiment, read all the steps of Techniques carefully. It contains information that may be helpful in designing your experiment.

POPULATION STUDIES

If you wish to observe what kinds of organisms are living in your milk now and during the course of your investigation you will probably want to grow them on agar plates. The procedure for preparing the plates, often termed “streaking,” enables you to examine more carefully what organisms are present at any point in time by amplifying their numbers. To do this:

- Obtain a nutrient agar plate and a cotton swab or inoculating loop. Place a nutrient agar plate on the table. Be careful not to touch the cotton end with your fingers.
- Dip the cotton end into the milk.
- Remove the lid of the agar plate and continue to hold the lid in your hand, being careful not to touch the inside.
- Gently rub or streak the cotton swab or inoculating loop across the surface of the agar. Do not press too hard. It is best to move the swab in a zigzag pattern as shown in Figure 8.1; you may want to streak your initials in script.

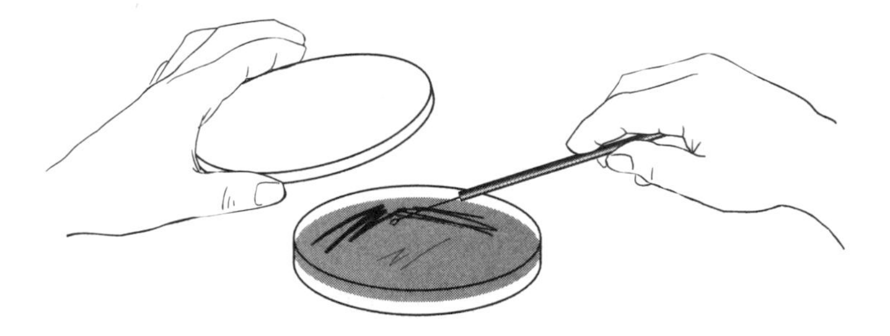

Figure 8.1
Streaking a plate

- Replace the lid on the agar plate and tape it closed. Label the plate on the bottom (using a wax marking pencil) with your group name, the date, and the time. Place the agar plate upside down (agar side up) in a 37°C incubator or in the warmest spot in the room.
- After a few days, check your agar plate for growth. Record any changes using an illustration.

The agar in the plate contains nutrients that many kinds of bacteria can use to grow. It also contains a substance called agar which is obtained from seaweed. Agar gives the nutrient mix a firmness that enables you to spread the organisms out on a surface and work with them. The procedure of “streaking” out your sample is actually a technique that permits separation of the different organisms that will grow on this plate. By spreading the sample across the large surface area of the plate you can obtain isolated growths of microorganisms called colonies.

MICROSCOPIC POPULATION STUDIES

If you wish to take a careful look and characterize the different kinds of organisms that might grow in your milk, you will probably want to examine samples taken from the agar plates under the microscope. Most bacteria are very small and difficult to see. Different stains will stain different kinds of bacteria and help in the identification process. For the purposes of this investigation, you will use one stain. When you take samples for viewing under the microscope, you may wish to sample from different-looking colonies. This will allow you to examine the different kinds of growth that may be occurring in your environment. The following provides a procedure for staining each sample.

PREPARE A MICROBIAL SMEAR

- Place a small drop of water on a microscope slide.
- Remove a small amount of growth from your agar plate using the wooden end of a cotton swab and mix it with the drop of water. You may wish to prepare several smears from colonies on your plate that look different.
- Air dry the slide for a few minutes. When it has dried pass the slide briefly through a flame (this is called heat fixing). Be sure the side of the slide with the specimen is facing up, away from the flame, when fixing.

CAUTION: *Keep hair and clothing away from flame.*

STAIN THE SMEAR

- Place several drops of crystal violet stain on the heat-fixed smear and wait two minutes.
- Rinse distilled water over slide and blot dry gently with a paper towel.
- Examine your slide under a microscope. Start with low power, then increase magnification. Draw what you see.

MEASURING pH

You may wish to measure the pH changes that occur in your environment. The pH scale found on pH paper or on a pH meter is a range of numbers running from 1 to 14. These numbers indicate whether a solution is acidic (pH 1 to around 6), neutral (pH 7.0) or alkaline (basic) (pH 8 to 14).

What determines whether a solution is acidic, neutral, or alkaline? Water can dissociate into the ions H^+ and OH^-; this means that the hydrogen has given up one electron to the hydroxyl (OH) group, leaving the hydrogen with a positive charge and the hydroxyl with a negative charge. Charged atoms are called ions. When these ions are present in equal amounts the pH is neutral. When a solution contains more H^+ ions than OH^- ions, it is acidic; when it contains more OH^- than H^+ it will be basic. A solution of water is usually around pH 6.5 to 7.5. If hydrochloric acid is added to the water, it will dissociate into its ions H^+ and Cl^-; the number of H^+ ions will now be greater than the number of OH^- ions and the solution will be acidic. If sodium hydroxide is added, it will dissociate into Na^+ and OH^- and the solution will be basic. The majority of biological reactions take place between 6.5 and 7.5. Measure the pH using litmus paper (the paper turns pink in acidic solutions and blue in basic solutions), pH paper, or a pH probe. Changes in odor are often indicative of changes in pH.

CONTINUING TO REAP THE SPOILS

[SUGGESTED DAYS 4, 7, 10]

PROCEDURE

1. Get your milk sample in the beaker from its storage (incubating) location.
2. Take careful notes describing the conditions of the area where your sample has been incubating.
3. After you have sampled and measured what you wish to measure and recorded all of your observations, cover the milk sample container with a piece of clear plastic wrap that you have pierced with the tongs of a fork or other pointed tool to aerate the container.
4. Return your milk sample to the designated storage location.
5. If you have agar plates from previous sessions, check them for growth. If they have grown and you are observing the different organisms, prepare and stain your samples and make your observations.
6. If you wish to keep your bacterial smears keep them stored inverted in a safe place.
7. In addition to recording your data, record any questions you may have or any additional unusual observations you make.
8. Think about the following questions as a guide to analyzing your results and thinking about what has happened to the milk and why:
   - What do the changes you have observed suggest about what has happened in the milk over the last several days?
   - How do these changes relate to metabolic processes that have occurred in organisms growing in the milk? Explain each change.
   - If you observed a change in the kind of microorganisms growing in the milk over time, why do you think this occurred?
   - Do you think the nutrient composition of the milk has changed? How? How would you go about verifying this?

EXTENDING IDEAS

ON THE JOB

MEDICAL LABORATORY TECHNICIAN Are you interested in using your organizational and detail-oriented skills working in the medical field? Medical lab technicians conduct routine tests in medical laboratories which help to diagnose and treat disease. A laboratory technician might be a generalist and carry out a variety of lab procedures, or might specialize and run lab procedures specific to a field such as cytology (study of cells), hematology (study of blood), serology (study of body fluids) or histology (study of body tissues). Technicians might prepare samples of body tissue or run lab tests such as urinalysis, blood counts, or chemical and biological analyses of cells, tissues, and other bodily specimens. Technicians use equipment such as blood analyzers which detect cholesterol, sugar, or hemoglobin levels in blood, microscopes to view tissue samples and sometimes electron microscopes. This information is used by physicians, surgeons, or pathologists (specialists in disease) to diagnose and treat patients. Lab technicians usually work in a hospital or in a lab affiliated with medical practices. Medical lab technicians might have either a high school diploma with one or two years of post high school education in a medical lab technician training program or a two year college degree in medical lab technician training. With a four year college degree advancement to a medical technologist is possible. This position means working on specialized or more complex work and sometimes supervising other lab technicians. All lab technicians are certified. Classes such as biology, chemistry, math, computer science, and English are necessary.

MICROBIOLOGIST Does the seemingly invisible world that appears under a microscope fascinate you? Microbiologists are scientists who study microscopic life. A microbiologist studies organisms such as bacteria, viruses, algae, mold, yeast, or any other organism that is of microscopic size. Microbiologists investigate problems such as studying organisms that cause disease in order to learn how the organism does it, or studying the microorganisms that cause pollution, or working on how to produce new vaccines against disease. A microbiologist might also work studying food science, life on other planets, agriculture, or microorganisms involved in environmental recycling. The lab techniques and procedures used by a microbiologist are not unique to microbiology, but are the techniques and equipment of chemists, geneticists, ecologists, or physiologists. With a four year college degree, positions as laboratory assistants, or clinical or research microbiologists are possible. With a master's degree or doctoral degree, microbiologists can teach in a university or pursue independent research. Classes such as biology, chemistry, physics, mathematics, computer science, and English are recommended.

How Did It All Begin?

PROLOGUE How did life begin? Did life arise from nonlife? How did the components of the molecules of life get organized into biomolecules and these biomolecules into cells? In the quest for answering the question "What is Life?" the question of where and how it all began inevitably follows.

Attempts to answer the question of how life began have been made through pursuits in the philosophy, religion, and science of every culture. The answers have inspired the creation of countless stories, poems, paintings, and musical scores. Ancient Egyptians thought that silt brought in by floods from the Nile gave rise to frogs and toads; the Greek philosopher Aristotle identified slime and drops of dew as the origin of insects, and mud as giving rise to fish and eels.

Speculation about the events which brought about that transition from nonliving to living has resulted in many heated debates. Since no living thing existed to witness the beginnings of life, scientists rely on fossil evidence and on studies of the living descendants of the first life—life on Earth today—to help unravel this mystery. Scientific explanations for the origins of life have been proposed, discarded, and continue to change and evolve as evidence and data are gathered from many different fields including oceanography, molecular biology, geochemistry, paleobiology, and astronomy.

Some theories propose that the transition from nonlife to life was a slow and measured event. In 1871, Charles Darwin proposed that life arose gradually over time in a "warm little pond." From a rich broth of organic chemicals, molecules coalesced (joined) to form the first stirrings of organisms. Other scientists have theorized that life exploded on the scene, suddenly and in many different places, in a bubbling cauldron, a boiling kettle of soup that contained all the necessary ingredients of life.

Another source of scientific exploration is to attempt to reproduce in the laboratory the actual beginnings. In 1953, Stanley Miller and Harold Urey recreated in a glass jar conditions that were believed to be present on Earth during its early formation: they used water (as oceans) and heat to form water vapor; methane (CH_4), ammonia (NH_3), and hydrogen for atmosphere; and electrical sparks for the lightning and electrical discharges that probably punctuated the atmosphere. Within a week, a sticky sludge was collected in a beaker (see Figure 9.1). When

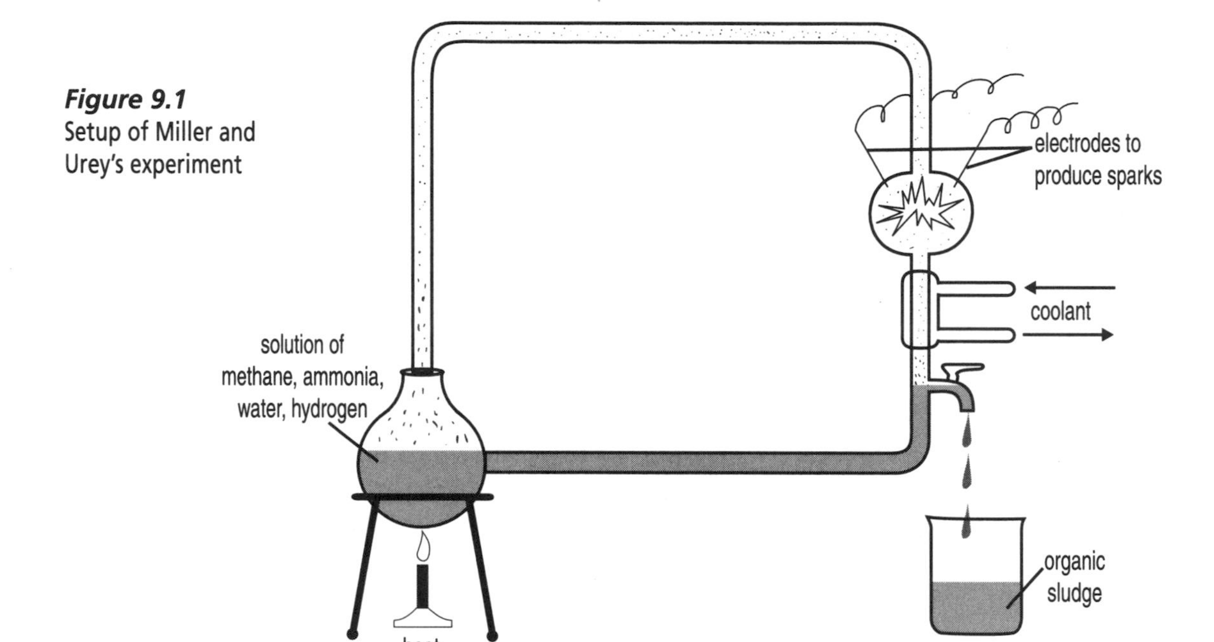

Figure 9.1
Setup of Miller and Urey's experiment

analyzed, this sludge proved to contain organic molecules, including large quantities of amino acids. Using similar conditions, scientists have been able to generate nucleotides, lipids, carbohydrates, and ATP. Were these conditions the ones that prevailed during the formation of the earth? Was this how the molecules of life were formed four billion years ago?

The Bubbling Cauldron

INTRODUCTION Condensing from a cloud of dust and gases, the earth was formed approximately 4.8 billion (4.8 x 109) years ago. Heat generated from the pressure of gravity, which pulled the earth into a spherical shape, and from the radioactive elements in the dust may have caused the interior of the earth to melt. Volcanic eruptions wracked the surface of the earth; the molten rock that spewed forth may have caused the atmospheric temperatures to reach fiery levels. The gravitational pull of the earth undoubtedly attracted meteors and asteroids from a heavens filled with celestial debris, remnants of the Big Bang which created the universe.

Although no one is sure of the composition of the air at that time, recent evidence suggests that instead of the atmosphere envisioned by Miller and Urey, the air over the surface of the earth was most likely made up of water vapor, carbon monoxide, carbon dioxide, nitrogen, hydrogen sulfide, and hydrogen cyanide (HCN). An inhospitable sphere bombarded by extraterrestrial missiles and having an atmosphere of cyanide and carbon monoxide does not sound exactly inviting for any form of life. Yet, from this inhospitable environment life apparently did arise.

TASK

1. With your partner, use your understandings of biomolecules to create a recipe for this "primordial soup" that would yield biomolecules.
2. Remember, you are limited to the materials and energy that were present on the earth at the time that it is believed life began, about 4.4 to 4.8 billion years ago.
3. As you formulate your recipe reread the Prologue to this learning experience. Include in your recipe the following:
   - what elements are required;
   - how the elements will fit together;
   - what is needed to form the bonds between the atoms; where this might come from;
   - what the first biomolecules might have looked like; and
   - how larger molecules might have formed out of smaller molecules.

THE SPARK OF LIFE

MASSES OF MOLECULES

Did life arise in the calm of the pond, the warmth of the pool, slowly and steadily over millions of years, as Darwin proposed? Or, did it arise frothing and spewing from a tumultuous, steaming environment? Recent discoveries near hydrothermal vents, cracks leading to chambers of molten rock beneath the ocean floor, have revealed ecosystems teeming with life and have lent support to the idea of a fiery origin of life.

Mounting evidence indicates that life most likely began earlier than previously thought, before Earth began to cool and develop a more hospitable atmosphere.

Other evidence suggests that the beginnings of life were brought to Earth by comets, asteroids, and meteorites colliding with Earth, or by the interplanetary dust that coated the newly forming planet. The meteorite that fell on Murchison, Australia in 1969, was rich in an array of carbon-containing compounds including several different amino acids, some of which have never been seen in organisms on Earth.

What did early biomolecules look like? Did they change over time? As you have investigated in the last several learning experiences, all living things consist of primarily four kinds of biomolecules. Does this mean these were the only molecules that formed during this early period? Or is it possible that many arrangements of molecules came together but these were the only ones that, for whatever reasons, "survived" and were well suited for becoming "life"? This last idea reflects one of the great unifying principles of biology, evolution. *Evolution* is the process by which changes occur in living things over time. While these changes occur at random, certain of these changes persist because they offer some selective advantage in survival and for persisting in a population. The biomolecules that persist today in all living things are the descendants of those early molecules and are the result of changes that occurred in the structure of biomolecules over time that made them well suited for survival. Figure 9.2 illustrates a possible timeline of events as they might have occurred over the course of the last 4.8 billion years.

Figure 9.2
Timeline of the history of life

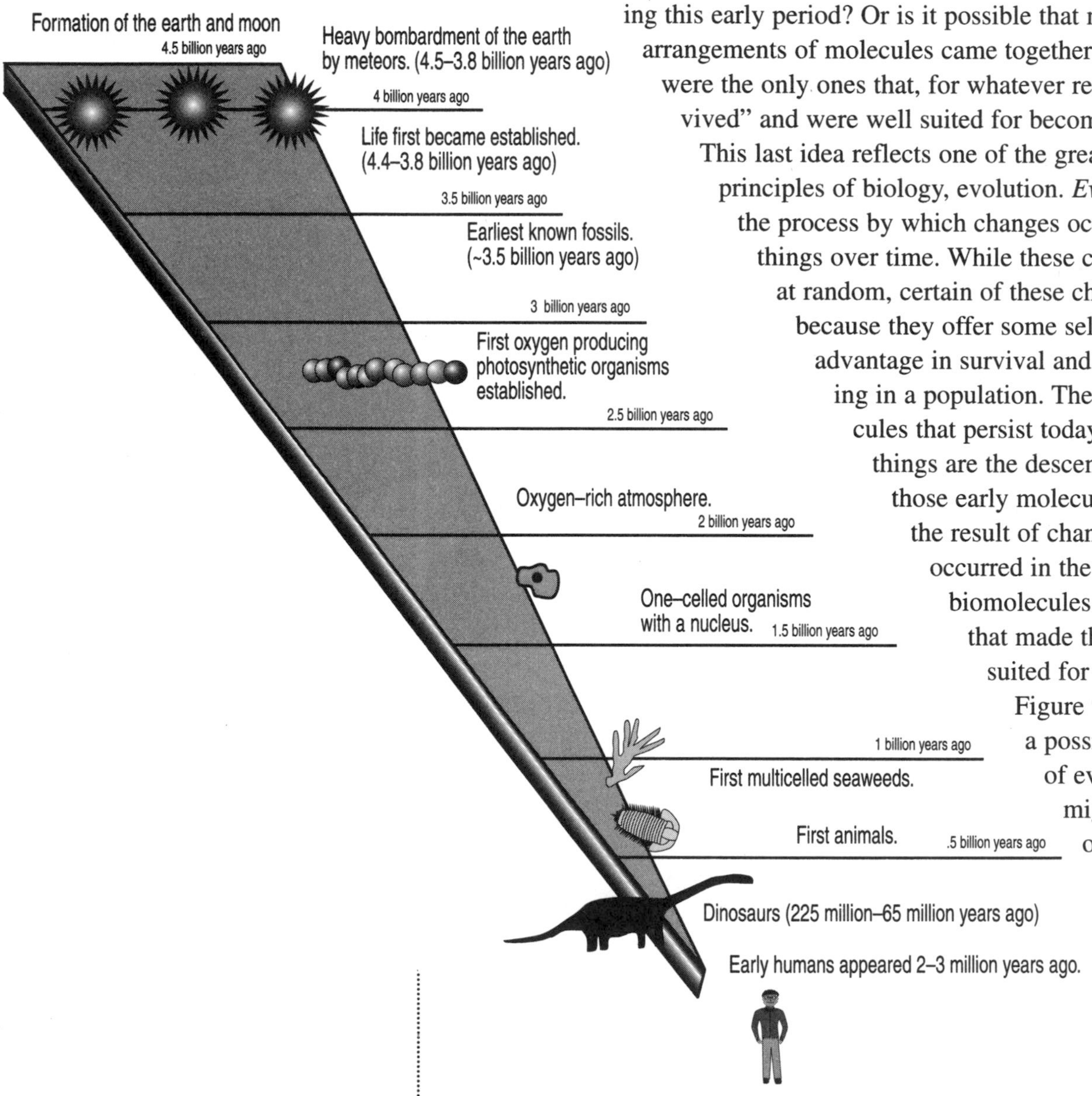

THE IMPORTANCE OF BEING ORGANIZED

Given that biomolecules were formed and accumulated in a rich organic broth, the question which arises is how and why did they organize into larger structures? How did these biomolecules ultimately secede from their environment to form their own, separate "union"?

In order for chemical reactions to occur efficiently, the reacting components must be in close proximity to one another. A chemist is more likely to carry out a reaction in a test tube than in a gallon vat or on a table top. Similarly, if life is dependent upon a series of chemical reactions, the molecules of life must be close together. The next step on the road to living organisms was the encapsulation of the large molecules of life in a kind of living "test tube."

In 1923, a Russian scientist, Alexander Oparin, proposed that early organic molecules did not remain as simple compounds but rather that they formed large aggregates or clusters as a way of becoming more stable. Very early precursors of cells might then have been formed when aggregates of large molecules were surrounded by a membrane-like structure. In 1960, American scientist Sidney Fox demonstrated experimentally that large clusters of proteins or ribonucleic acids, when heated and allowed to cool under appropriate conditions, would form droplets or microspheres. These microspheres were able to grow, reproduce, and perform some of the basic life processes such as breaking down glucose. Lipid molecules are likely candidates for this membranous kind of housing. Several theories exist about how biomolecules might have become encased in a lipid membrane. Experiments have shown that when a broth of concentrated organic molecules is dried, lipids spontaneously form droplets. It has also been shown that this drying process results in the chemical bonding of ribonucleotides to form large molecules of RNA. Could the association of these ribonucleotide molecules in primordial soup have led to the formation of membrane-bound ribonucleic acids?

In another model, proposed recently by biologist David Deamer, molecules that formed membranes were also introduced by meteors. Deamer extracted organic compounds from the meteorite that had landed in Murchison, Australia, and demonstrated that among these compounds was a fatty acid chain, nine carbons long, which was capable of forming a membranous sack, a *liposome*. Further research with liposomes made in the laboratory has shown that these structures could take up RNA.

What is there about RNA that could make it a candidate for a component in the earliest life-form? RNA is the nucleic acid that, in most organisms today, carries the information from DNA to the site in the cell where protein is synthesized. Some life-forms use RNA instead of DNA for information storage. One property of RNA is that it is highly changeable, that is, the information carried in RNA can change more easily than that in DNA. Another property of RNA is the ability to replicate or make copies of itself. The ability to make copies of oneself is a distinct advantage—the more copies, the more opportunity for change. The more opportunity for change, the more likely it is that some variant of the original molecule (or organism) will be better suited for survival.

Could RNA in a lipid shell have been the first cell-like structure? The association of self-replicating RNA inside a lipid membrane shell may not seem like much of a life. But it may have marked the beginnings of a line of cell-like structures that, through the process of change over time, evolution, ultimately resulted in the structure of the cell as we know it. In the next learning experience you will be investigating the cell and its complex structures which enable it to carry out the biological processes of life.

ANALYSIS

Write responses to the following questions in your notebook.

1. Examine Figure 9.2. What major events are indicated that are important in providing evidence for the origins of life? What is the significance of these events?
2. What events in the timeline and in your understanding of living organisms suggest that all life may have descended from common origins?
3. Could the droplets created by Sidney Fox be considered living? Could David Deamer's RNA-containing liposomes? Use your understanding of the characteristics of life to explain your response.
4. A well-known biologist once stated: "Given enough time, anything is possible." Explain what this means in the context of the origin of life.
5. Create a concept map that describes how life is thought to have begun, starting with primitive Earth and the origins of life based on the readings. Include all of the components that you consider essential, how these might have organized into the biomolecules and how the biomolecules might have come together into a primitive organizational structure or cell.

EXTENDING IDEAS

ON THE JOB

OCEANOGRAPHER Would you be interested in spending time outdoors on the ocean? Oceanographers study the physical, chemical, geological, and biological compositions of the ocean and the sea floor. There are four main branches of oceanography and although an individual might focus on one branch, there is tremendous overlap which means that the individual is familiar with all branches of the field. Biological oceanographers, also known as marine biologists, study all aspects of the plant and animal life found in the ocean. Physical oceanographers study the physical aspects of the ocean such as water temperature, water density, the motions of ocean water (waves, water currents, and tides) and the relationship between the ocean and the atmosphere. Geological oceanographers study the physical composition and the topographical features of the ocean floor. Chemical oceanographers study the chemical composition of ocean water and the ocean floor. Oceanographers collect information about the ocean through observations, surveys and experiments completed using specially designed equipment. An oceanographer might use equipment that uses sound waves to measure depth, special thermometers to measure temperature, deep sea diving gear for underwater diving, submarines to explore deep water and, possibly in the future, satellite sensors. A four year college degree is required for laboratory or research assistant positions in oceanography and a master's or doctorate is needed to teach in a university or to pursue independent research. Classes such as physics, chemistry, biology, mathematics, oceanography, foreign languages and English are useful.

Night of the Living Cell

PROLOGUE In the past several learning experiences you have been exploring the biochemical basis of life. You have examined how simple molecules such as carbon dioxide and water can be transformed into more complex molecules and how complex molecules can be broken down and rearranged by metabolic processes to produce a vast diversity of complex molecules which make up living organisms.

In essence, you have been examining the molecular architecture of life—how simple components can be combined to make complex and intricate structures. In this learning experience you will examine the next step in this organizational hierarchy, the cell. All of the components of a cell are composed of the biomolecules you have been examining. These biomolecules are organized into structures which have specific functions in the cell and are the sites in the cell where the metabolic processes you have been studying take place.

Our Body, Our Cells

Organisms, from the simplest to the most complex (including humans), are composites of cells. This is not a very flattering image at first—we like to think of ourselves as highly evolved and complicated organisms, finely tuned to the business of living. On some levels we are, but the ways in which our cells function and interact with one another is what makes being human, or plant, or yeast, possible.

The human body has several trillion cells. Many of the fundamental processes of life which enable organisms to live are carried out at the cellular level. Cells take in resources such as nutrients, water, and gases from their environments. Cells use these resources to transform energy and to synthesize biomolecules that can be used to build new components of the cell. Despite the enormous diversity we see among animals, plants, and bacteria, the functions and chemical composition of all cells are remarkably similar.

Cells themselves may be the most complex units around—more complex in many ways than the bodies of which they are part. The cell carries out many different activities and coordinates the complex web of chemical reactions which make "life" happen. The work of all cells, like the work of the living organisms that you have investigated so far, includes taking in nutrients from resources in the environment, then breaking them down and reconstructing them in different ways. Individual cells also respond to their environment, repair and maintain themselves, and replicate themselves—all the characteristics of life that you have been exploring.

The Whole Cell and Nothing But the Cell

INTRODUCTION What is in the cell that allows it to carry out metabolic processes? In the following activity you are going to identify where in the cell these processes occur. You will research what structures are in a cell and determine the metabolic processes which take place within them. You will then determine how structures within the cell work to ensure that these processes occur efficiently and effectively. Then, using available materials, you will build a large model of your cell. When the model is completed, each group will make a presentation to the class, describing its cell.

MATERIALS NEEDED

For each student:

- 1 sheet of chart paper (optional)

For the class:

- assorted model-building materials such as:
  - pipe cleaners
  - drinking straws
  - plastic or wooden beads
  - polystyrene balls and other shapes

- pasta in a variety of shapes, sizes, colors
- clay in a variety of colors
- yarn
- construction paper
- cardboard sheets and boxes
- plastic or paper bags
- glue

- biology texts and other reference books

PROCEDURE

1. Choose which cell type you would like to build: either an animal, a plant, or a bacterial cell. You may wish to choose a specialized cell such as a nerve cell, a blood cell, or a leaf cell.
2. Prior to building the model of the cell, you should:
   - find pictures in textbooks;
   - determine the structures that are components of the cell;
   - determine a scale for your model (such as 1:1000) and the approximate scale for the structures within (optional);
   - identify what biomolecules make up the various components of your cell; and
   - choose materials that reflect the structure and function of the cell components. (You can be creative and use your own materials that are not on the list.)
3. Begin to build your group's model.
4. Discuss how the cell that you are constructing takes up nutrients and other essential resources and uses them to sustain life.
   - Outline the metabolic processes. Use any notes and readings from this module to assist in creating this outline.
   - Label your model organelles with the metabolic processes they perform.
   - Create a diagram which shows the structures (*organelles*) of the cell and their metabolic functions. How might the function of one organelle affect another organelle?
   - If you choose a specialized cell be sure to research any specialized structures or molecules that enable the cell to carry out its specialized functions.
5. Visit other groups as work progresses to observe their models, ask questions, and compare and contrast their models with that of your group.
6. Decide how your group will make its presentation. Split up the tasks so that each group member has a part to prepare for the pre-

sentation. Your group will have about 10–15 minutes to organize your presentation at the beginning of class. Each presentation will be given approximately five minutes. Your presentation must include the completed model, your diagram of the pathway that connects organelles through the metabolic process, and the following information:

- a description of the biomolecules that make up the various components of your cell;
- the structures involved in moving nutrients and other external resources from the outside environment of the cell to the inside of the cell and in removing waste products;
- a description of how biomolecules might move around in the cell;
- the relationships between breakdown and biosynthetic activities, where these activities occur, and how the products or intermediates might need to move to different places in the cell;
- the structures within the cell that are involved in obtaining energy and transferring it into a form the cell can use; and
- a description of how you think the organizational structure of the cell facilitates (makes easier) its ability to carry out the essential functions of life.

7. Present your model to the class.

As The Cell Turns

"The cell is the basic unit of life" is a definition that you may have heard at some time. What do you think it really means?

A CELL AND ITS RESOURCES

As you have been investigating, living organisms must take up resources, including nutrients, water, gases, and energy, from their environment in order to maintain life. What happens to these resources within an organism? The processes for chemical and energy transfer occur primarily at the cellular level, but some processes, such as digestion of nutrients begin outside the cell before other processes within cells can occur. Single-celled organisms break down substances in the environment while they are still outside the cell, using compounds that the organism secretes into its immediate environment. The nutrients are then absorbed through small holes (pores) in the surface of the cell or are engulfed by the cell (a wrap-around eating process).

Digestion of nutrients from food begins in multicellular animals when the animal ingests food. The breakdown process of complex carbohydrates starts in the mouth. Food continues to be digested in the stomach and the resulting products are passed on to the small intestine where further digestion occurs. By this time the original food—though it has not yet entered any cells—has been broken down into its basic components such as *sugars*, amino acids, and fatty acids. These components are then be absorbed into individual cells by mechanisms similar to those carried out by single-celled organisms.

Other resources, such as gases and water, are also taken into the organism. Little or no breakdown is necessary as these can be readily absorbed directly into cells for use. Plants also take in nutrients (in the form of minerals), gases, and water, and these resources eventually make their way to the cells of the plant.

YOU ARE WHAT YOU METABOLIZE

Whatever the organism is and however the above-mentioned resources reach its cells, once those resources have entered the cell, the metabolic processes occur much in the way that you have been studying. A cell, whether bacterial, plant, or animal, takes the resources from its environment and transforms (metabolizes) them into materials and energy it can use. Animal and bacterial cells degrade complex biomolecules, break them into building blocks, and transfer the chemical energy from the environment into a form of chemical energy they can use, as you determined in Learning Experience 7—A Breath of Fresh Air. Plant cells take in simpler molecules, capture light energy from the sun and, using photosynthesis, transform these resources and light energy into materials the cells can use. Using this energy and these building blocks, all living organisms then synthesize the biomolecules every cell can use to build new parts of itself and to carry out the processes of life.

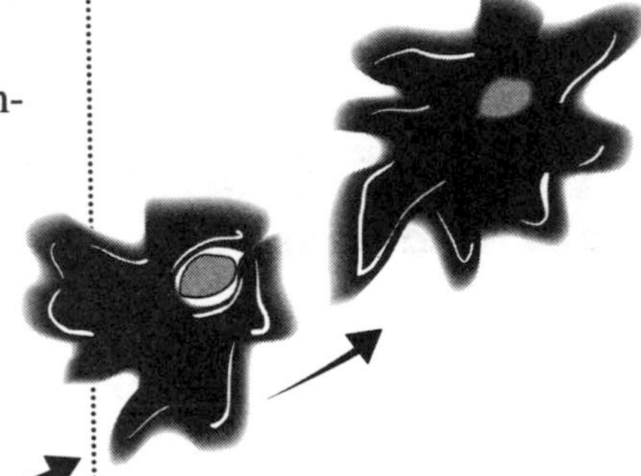

Figure 10.1
Engulfment (endocytosis) by a cell

Cells are composed of biomolecules: proteins, lipids, carbohydrates, and nucleic acids. These complex chemical compounds join together with other biomolecules to form structures which make up the cell. Each of these molecules has chemical properties which are suited to its role in the cell. What roles do these biomolecules play in the cell? As you have seen, carbohydrates are used by the cell as an energy source, as structural support, and of course, as starting materials for the synthesis of biomolecules. Fats store energy and provide the materials for important structural components of the cell such as membranes. Nucleic acids are used to store and transmit the information in cells. Proteins are the main workhorses of the cell; they form important structural components of the cell and provide energy. In addition, enzymes, which are proteins, are responsible for facilitating all of the life processes you have been examining. The breakdown of complex molecules, the release and capture of energy in new chemical forms, and the synthesis of new biomolecules from the building blocks—all are carried out by proteins.

Metabolism is a very dynamic activity. It is the capacity of cells to get energy and utilize it to build, break down, store usable substances, and get rid of waste substances, all in a controlled way. Metabolic processes make available the molecules which are required for the cell to make new parts for itself, to repair itself when damaged, to communicate with its environment and (in the case of multicellular organisms) with other cells, and ultimately to replicate itself. These processes are occurring at every moment in the life of the cell, and are multiplied by the trillions of cells in your body, even as you read this!

READING

An Organelle With A View

One of the differences between different types of cells is the presence of structures within plant and animal cells which are absent in bacterial cells. The space inside plant and animal cells is divided by membranes into complicated structures known as organelles, each of which carries out a portion of the interconnected processes of metabolism. These structures give organization within the cell.

Another difference between bacterial cells (also called *prokaryotes*) and plant and animal cells (called *eukaryotes*), which you may or may not be aware of (depending on whether you created your model to scale or not), is the vast difference in size. Plant and animal cells are about one thousand times bigger than the average bacterial cell. How do you think these differences in structure within cells and cell size might be related?

The following excerpt describes the differences between prokaryotic cells and eukaryotic cells; why there came to be differences; and why organelles contribute to the efficient functioning of eukaryotic cells.

As you read, draw diagrams, create a table, or create separate concept maps of a prokaryotic cell and a eukaryotic cell. Compare the three main structural aspects that differentiate prokaryotic cells from eukaryotic cells.

> *Like most inventions, life started out simple and grew more complex with time. For their first three billion years on earth, living creatures were no larger than a single cell [prokaryotes]. Gradually, the forces of natural selection worked on these sim-*

ple organisms until eventually they became bigger, more sophisticated and more intricate. Organisms increased in size not only because the individual cells grew but also because multiple cells—in some cases many millions—came together to form a cohesive whole. The crucial event in this transition was the emergence of a new cell type—the eukaryote. The eukaryote had structural features that allowed it to communicate better than did existing cells with the environment and with other cells, features that paved the way for cellular aggregation and multicellular life. In contrast, the more primitive prokaryotes were less well equipped for intercellular communication and could not readily organize into multicellular organisms. . .

Not only do eukaryotic cells allow larger and more complex organisms to be made, but they are themselves larger and more complex than prokaryotic cells. Whether eukaryotic cells live singly or as part of a multi-cellular organism, their activities can be much more complex and diversified than those of their prokaryotic counterparts. In prokaryotes, all internal cellular events take place within a single compartment, the cytoplasm. Eukaryotes contain many subcellular compartments called organelles. Even single-celled eukaryotes can display remarkable complexity of [structure and] function . . .

On a very fundamental level, eukaryotes and prokaryotes are similar. They share many aspects of their basic chemistry, physiology, and metabolism. Both cell types are constructed of and use similar kinds of molecules and macromolecules to accomplish their cellular work. In both, for example, membranes are constructed mainly of fatty substances called lipids, and molecules that perform the cell's biological and mechanical work are called proteins. . . Both types of cells use the same bricks and mortar, but the structures they build with these materials vary drastically.

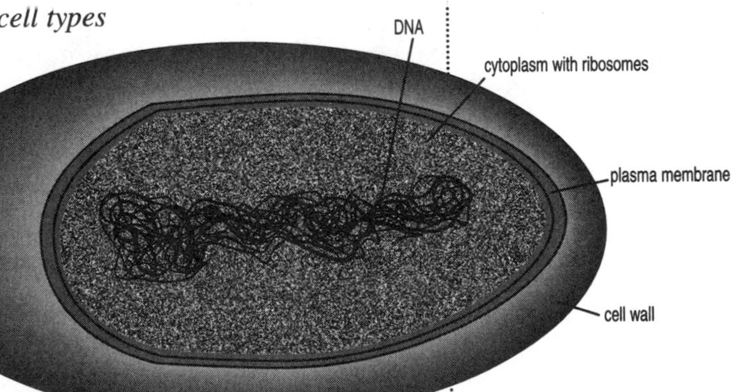

Figure 10.2
A prokaryotic cell
(Not drawn to scale.)

The prokaryotic cell can be compared to a studio apartment: a one-room living space that has a kitchen area abutting the living room ,which converts into a bedroom at night. All necessary items fit into their own locations in the one room. There is an everyday, washable rug. Room temperature is com-

fortable—not too hot, not too cold. Conditions are adequate for everything that must occur in the apartment, but not optimal for any specific activity.

In a similar way, all of the prokaryote's functions fit into a single compartment. The DNA is attached to the cell's membrane. [Structures for synthesizing proteins] float freely in the single compartment. Cellular respiration is carried out at the cell membrane; there is no dedicated compartment for respiration.

A eukaryotic cell can be compared to a mansion, where specific rooms are designed for particular activities. The mansion is more diverse in the activities it supports than the studio apartment. It can accommodate overnight guests comfortably and support social activities for the adults in the living room or dining room, for children in the playroom. The baby's room is warm and furnished with bright colors and a soft carpet. The kitchen has a stove, a refrigerator and a tile floor. Items are kept in the room that is most appropriate for them, under conditions ideal for the activities in that specific room. [However, items from one room may be needed in another room for the functions of both to be carried out; for example, food prepared in the kitchen must be carried to the dining room in order to be consumed, and the waste generated in the kitchen must be removed to the trash cans outside.]

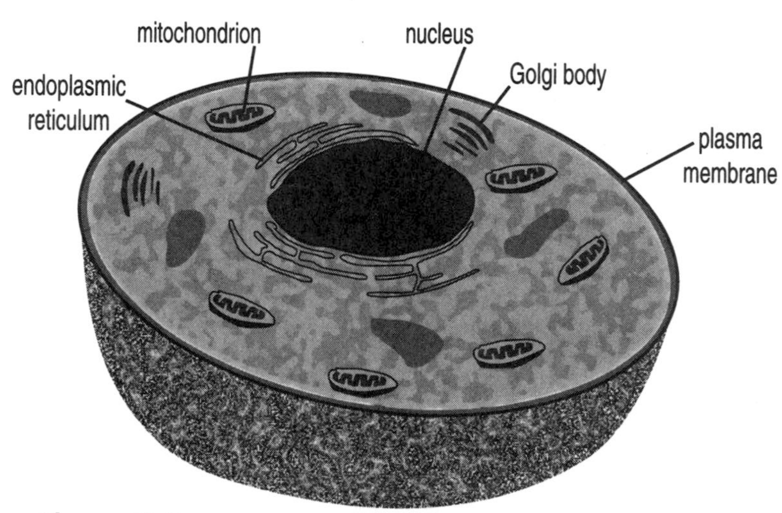

Figure 10.3
A eukaryotic cell

A eukaryotic cell resembles a mansion in that it is subdivided into many compartments. Each compartment is furnished with items and conditions suitable for a specific function, yet the compartments work together to allow the cell to maintain itself, to replicate [itself] and to perform more specialized activities.

Taking a closer look, we find three main structural aspects that differentiate prokaryotes from eukaryotes. The definitive difference is the presence of a true (eu) nucleus (karyon) in the eukaryotic cell. The nucleus [separates] the DNA in its own compartment... from the rest of the cell. In contrast, no such housing is provided for the DNA of a prokaryote. Instead, the genetic material is tethered to the cell membrane and is otherwise allowed to float freely in the cell's interior....

The organelles of eukaryotes include membrane-bounded

compartments such as the lysosome, a highly acidic compartment in which digestive enzymes break down food. The endoplasmic reticulum is an interconnected system of membranes in which lipids are synthesized. ...[In] another membrane system called the golgi apparatus...proteins are... [transported to other places in the cell or to the outside of the cell]. Eukaryotic cells contain special energy centers. In animal cells these are the mitochondria; plant cells have chloroplasts as well as mitochondria. Within mitochondria, organic compounds are broken down to generate the energy-rich molecules [which] provide energy for the cell's biochemical reactions. [Prokaryotes do not have organelles, but some types have infoldings within their plasma membrane where some metabolic reactions may occur. Certain of these infoldings may have at some point extended so far into the cell's interior that they became channels to the surface of the cell. Perhaps some of these evolved into separate compartments that provided protection for certain components from foreign or harmful substances.]

The third distinguishing feature between the two cell types is the way in which the cell maintains its shape. Cells...have skeletons [plasma membranes] and...the cellular skeleton can be either internal or external. Prokaryotes have an external skeleton; a strong wall of cross-linked sugar and protein molecules surrounds the cell membrane and is made rigid by the [water] pressure of the cell. The wall lends structural support...and...helps to maintain a barrier between substances inside and outside the cell. Such an external skeleton...limits communications between cells....

The skeleton of the eukaryotic cell is internal; it is formed by a complex of protein tubules....The internal placement of the cytoskeleton means that the surface exposed to the environment is a pliable membrane rather than a rigid cell wall. The combination of an internal framework and a nonrigid outer membrane expands the repertory of motions and activity of the eukaryotic cell [and permits the cell greater communication with its environment and with other cells, which is a function of certain proteins].

—An excerpt from K. Kabnick and D. Peattie. "Giardia: A Missing Link Between Prokaryotes and Eukaryotes." American Scientist *79:34-43, 1990.*

Eukaryotic cells need their compartmentation because they are huge. A molecule drifting around inside a bacterial cell will sooner or later meet something suitable with which to react. In a eukaryotic cell it could drift for its entire life. By walling off compartments, a larger cell keeps control over its content. It also provides the potential for diversity of function which enables eukaryotic cells to come together, specialize, and form multicellular organisms.

ANALYSIS

Write responses to the following questions in your notebook:

1. What structures does a eukaryote have that a prokaryote lacks? What structures do they have in common?
2. How do metabolic processes occur in each cell type? Where do these processes occur in eukaryotes? In prokaryotes?
3. Describe advantages and disadvantages of eukaryotes. Then describe the advantages and disadvantages of prokaryotes based on the diagrams, table, or concept maps you created while reading "Giardia: A Missing Link Between Prokaryotes and Eukaryotes."
4. "Organisms are not larger because they are more complex; they are more complex because they are larger."—J.B.S. Haldane. Using the information from the reading "An Organelle With a View," and any other ideas or concepts you have thought about during your research on cells, write a paragraph which supports this statement.

EXTENDING IDEAS

- How did eukaryotic cells get organized into compartments? Did cell membranes come together as a way of being more efficient, or are compartmentalized cells an example of cooperation among different cell types? Research the theory which suggests that organelles derived from prokaryotic organisms which got "organized"—mitochondria from bacteria, and chloroplasts from blue-green algae. Describe the theory and then defend or refute it.
- Do you own the cells in your body, or can someone use them to make a profit and not share the rewards with you? Read the magazine article "Cells for Sale" (by Judith Stone, *Discover*, August 1988, pp. 33–39) and then write your opinion about whether John Moore should reap some profit from the use of his cells by a pharmaceutical company. Justify your decision using information from the article and any other resources you may have.

ON THE JOB

PHARMACOLOGIST Have you ever wondered how your doctor knows which and how much medicine to prescribe? Pharmacologists conduct basic and applied medical and drug research to study the effects of drugs and chemicals in animals and humans. One area of pharmacology studies how the chemical structure of the substance interacts at the cellular and molecular level in organs and tissues. This research results in standardizing drug dosages, and discovering how drugs should most effectively be used. Pharmacologists were involved in the development of anesthetics, antibiotics (such as penicillin), vaccines (such as tetanus and polio), tranquilizers, and vitamins. Pharmacologists may specialize in the effects of drugs on a particular part of the body such as the nervous system. A second area of pharmacology researches substances used in the environment, agriculture, or industry and specifically studies chemicals, food additives and preservatives, pollutants, poisons, and other materials to determine their effects on humans. Pharmacology is a different career from pharmacy. Pharmacy is the health profession which prepares and dispenses drugs to patients. Pharmacologists have a Ph.D. in pharmacology from a medical school or school of pharmacy and sometimes are also medical doctors (M.D.). Classes such as chemistry, biology, physiology, mathematics, computer science, physics, and English are recommended.

MARKETING, SALES OR CUSTOMER SERVICE REPRESENTATIVES Are you interested in how science is used in a business environment? Positions in the marketing and sales divisions of health care or biotechnology companies call for individuals with a background in both science and business. Market research analysts are responsible for identifying the market for the company's products or services, identifying other companies with similar products (the competition), analyzing what makes the company and its products unique, and deciding how to advertise these products to the customer so that the customer will purchase them. The business background provides the tools to understand the marketplace and the science background provides the tools to understand the company's products and what makes these unique compared to the competitors' products. Sales representatives are responsible for the direct sales of the services or products of the company. They call prospective and current customers, provide product information, and may conduct product demonstrations. A background in business provides the tools to recruit new accounts and grow existing accounts, while a science background helps in understanding the products and the needs of the prospective customer. Customer service representatives are responsible for ensuring that customers receive products or that services meet customer requirements within manufacturing capabilities. Customer service representatives respond to customer inquiries and comments about satisfaction with the product, take orders for products, and investigate problems related to the shipment of products, credits, and new orders. A science background is useful in understanding the nature of customer comments. A four year college degree is necessary for any of these positions. Classes in biology, chemistry, English, computer science, and communications are recommended.

The Great Divide

PROLOGUE Growth has been identified as one of the characteristics of life, but what does "to grow" mean? *Growth* can be defined as an orderly increase of all the components of an organism, a process that is dependent on the efficient uptake of nutrients. The result of this growth is observable as an increase in the size of a living organism. Since living organisms are made up of cells, growth must be viewed as a process that occurs at the cellular level. Cells increase in size by taking in nutrients, breaking them down and using the resulting energy and building blocks to make more cellular components.

The maximum size that many cells can reach is limited. At a certain size they must replicate (or duplicate themselves); that is, they must divide. After a division the two new cells, called daughter cells, will continue growing until the size limit is reached, and again division will occur.

Why can cells only grow to a certain size? What is the limitation on cell size? Most of this limitation has to do with the cell's ability to access its life sustaining nutrients. As you saw in earlier learning experiences, the ability to move resources into and out of cells is essential for the cell to survive. Nutrients must enter and byproducts must be removed. In fact, the ability to obtain building blocks and energy from resources is one of the characteristics of life, and enables the other characteristics of life to be maintained.

In this learning experience, you will explore how another important life process, cell division, is also dependent on the process of metabolism. You will see how it enables other characteristics of life to take place, including the abilities of an organism to grow, to maintain and repair itself, and to replicate. In the following investigation, "Soaking It All In," you will relate cell size to the ability to obtain nutrients and then determine the relationship of cell division to the maintenance of the characteristics of life.

SOAKING IT ALL IN

INTRODUCTION Efficient access to nutrients plays a role in determining how large a cell is able to grow. Using agar cubes to model the cell, you will examine the process of diffusion of substances through the cell membrane. You worked with agar earlier, when you grew microorganisms on nutrient agar plates. The agar in this activity contains a chemical indicator similar to the phenol red you used in detecting CO_2 in Learning Experience 7—A Breath of Fresh Air. Phenolphthalein is a chemical indicator that changes color in the presence of a basic or alkaline solution. By placing this agar block in a basic solution of sodium hydroxide (NaOH), you will investigate the question "How does the size of a cell affect its ability to obtain nutrients?"

MATERIALS NEEDED

For each pair of students:

- 2 pairs safety goggles
- 2 pairs of disposable gloves (optional)
- 1 small block of phenolphthalein agar
- 1 petri dish
- solution of 0.1 molar (M) sodium hydroxide (NaOH)
- 1 beaker (250- or 500-mL)
- 1 ruler
- 1 scalpel or utility knife
- 1 spoon
- graph paper
- paper towels

PROCEDURE

SAFETY NOTE: *Always wear safety goggles when conducting experiments.*

1. **STOP & THINK** Make predictions with your partner about the following:

   – Which will take up more nutrients within a given period of time—a large cell or a small cell?
   – Which will have greater access to the nutrients it takes up—a large cell or a small cell?

   Record your predictions in your notebook.
2. Obtain a block of phenolphthalein agar from your teacher.

3. Carefully cut out three small agar cubes from this block. Make the cubes of much different sizes: one 3 cm on each side, one 1.5 cm on each side, one 0.5 cm on each side.
4. Place the three cubes in a beaker; be sure that the cubes do not touch. Cover them with a solution of 0.1 M NaOH.
5. Every 2 minutes for a 6-minute period, turn the cubes over with a spoon so that all sides of each cube become exposed to the solution.
6. After 6 minutes remove the cubes from the solution and place on a paper towel. Carefully pat the cubes dry with another paper towel.
7. Without touching the cubes with your hands, cut each cube in half with a scalpel or knife. It is important to make one clean cut for each cube. Be sure to wash your hands at the end of this experiment.
8. Measure in centimeters how far from the edge the color change has occurred in each cube. What is the cause of this color change?
9. Compare how far the NaOH has diffused into each cube by calculating the percentage of each block that has changed color.
   a. First determine the volume of each cube. Calculate the volume by multiplying the height (h) by the width (w) by the length (l). For example, in a cube 3 cm on each side: The volume of the cube would be 3 cm x 3 cm x 3 cm = 27 cm^3. (See Figure 11.1a) Calculate the volume of each of your cubes and record the volumes in your notebook.
   b. Determine the volume of the uncolored area by measuring one side of the uncolored area. For example: (see Figure 11.1b)

      In this example, one side of the uncolored area measures 2 cm. Therefore the volume of the uncolored area is 2 cm x 2 cm x 2 cm = 8 cm^3.

      Calculate and record the volume of the uncolored area of each of your cubes.
   c. Determine the volume of the colored area by subtracting the volume of the uncolored area from the total volume of your cube: In this example,

27 cm^3	–	8 cm^3	=	19 cm^3
total		uncolored		colored volume

      Calculate and record the volumes of the colored area of each of your cubes.
10. Determine how much diffusion occurred in the different cubes by calculating the percentage of each cube into

CAUTION: *Scalpels and knives are sharp. Cut the agar on a hard surface and cut away from yourself.*

CAUTION: *NaOH is caustic. Avoid contact with your skin. If it comes in contact with skin, rinse thoroughly with water. If you should get any in your eyes, irrigate them immediately and inform your teacher.*

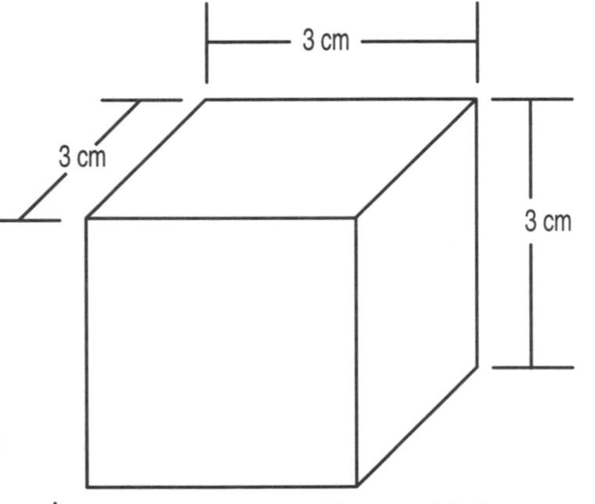

Figure 11.1a
Determine the volume of an agar cube

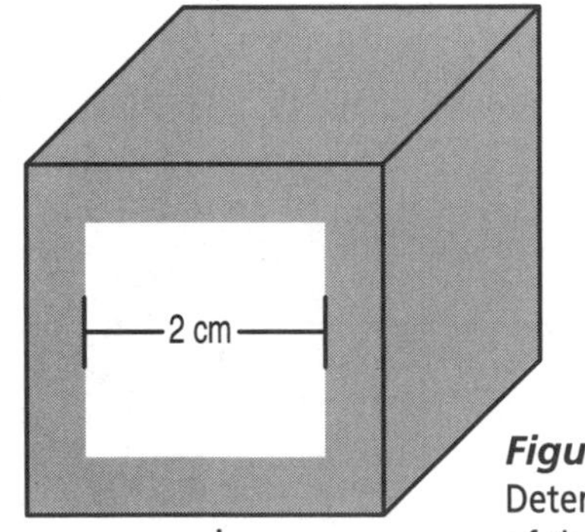

Figure 11.1b
Determine the volume of the uncolored area

which the NaOH solution has diffused. Divide the volume of the colored area by the volume of the total cube:

$$\frac{19 \text{ cm}^3}{27 \text{ cm}^3} = .704 \times 100\% = 70.4\%$$

Calculate and record the percentage of diffusion in each of your cubes.

11. Plot on a graph the percentage of each cube that the NaOH solution entered vs. the size of the cube (cm on one side).

ANALYSIS

Write responses to the following questions using your data:

1. Which cube had the greatest percentage of its volume reached by the NaOH solution?
2. How does the data compare to your predictions? If they differ, explain why.
3. If the agar blocks were cells and the NaOH solution were a nutrient solution required by the cell, which cell size would have the greatest advantage of nutrients diffusing into the cell? Why?
4. How might each of the agar blocks, as cells, get nutrients inside most efficiently?
5. Which size cell has an advantage for growth?
6. What might a cell do in order to change its surface-to-volume ratio, that is, to increase its efficiency in obtaining nutrients?

DIVIDE AND CONQUER

SERVICE IS SO SLOW IN THIS RESTAURANT

One of the characteristics of life that was identified in Learning Experience 1—Living Proof was the ability to grow. In multicellular organisms growth is generally associated with getting "bigger." Physical changes that occur during the transition in life between a young organism to a mature organism may involve significant increases in size. How does this growth occur? Growth is the result of events that occur at the cellular level. Individual cells increase in size as they use nutrients and energy to build new cellular components. These nutrients are taken into the cell through the cell membrane and are metabolized.

When the cell is small, areas in the cell where metabolic activity occurs are very near to the membrane through which the materials are being transported. However, as cells increase in size, the interior portions of the cell become further and further from the membrane. The distance from where nutrients enter the cell to where they are used increases and the functioning of the cell becomes inefficient. As in the agar block experiment, it takes much longer for substances to diffuse to the interior of a large cell than to a smaller one. Waste substances accumulate in the cells and must be moved from the interior of the cell where they are formed to outside the cell. The larger the cell is, the longer that takes, and the more energy is used.

For many types of cells, the method for dealing with this problem is to divide and form two smaller cells (*daughter cells*) out of one oversized cell. By dividing into two smaller cells, the large cell increases its surface-to-volume ratio and thus increases the efficiency with which substances can reach all cellular components in the interior of the cell. Visually compare the surface-to-volume ratios of the cells in Figure 11.2.

DIVIDING THE GOODS

The mechanism of cell division is a precisely orchestrated event that occurs in the same pattern in every eukaryotic cell (with some variation in cells that eventually form eggs and sperm cells). It requires that the parent cell 1) duplicate all the information that the daughter cells will need to carry out the exact same metabolic activities and life sustaining functions as the parent cell; 2) transfer this information to the daughter cells; and 3) provide the daughter cells with enough of the cellular machinery to be able to sustain themselves until they are making their own cellular parts and products. How does this happen?

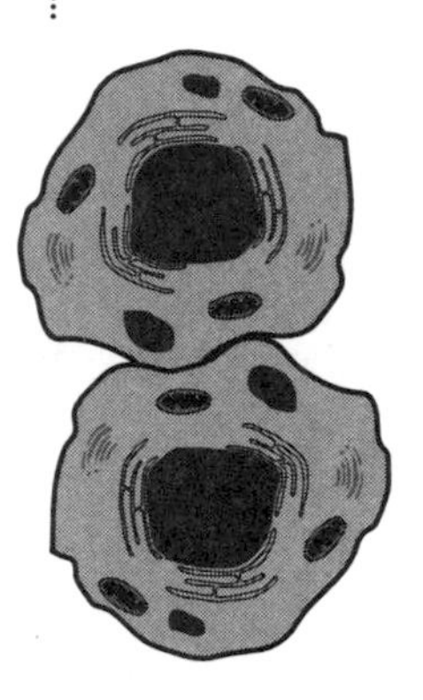

Figure 11.2
One large cell divides into two smaller cells. How does the surface-to-volume ratio change?

As described in Learning Experience 5—The Lego of Life, all of the information needed by a cell to carry out its functions is stored in its DNA. In eukaryotic cells the DNA in association with certain proteins is located in the nucleus within structures called chromosomes. The first step in cell division is to ensure that the chromosomes are faithfully duplicated, for it is these structures that will carry all the required information to the daughter cells to ensure that they are exactly like the parent cell.

The process begins with the replication of the DNA within each chromosome. The new copy of DNA forms associations with chromosomal proteins, resulting in two copies of the same chromosome. The chromosomes then sort themselves in such a way to ensure that each daughter cell receives one copy of each chromosome. Following chromosome duplication and separation, cell division occurs and two daughter cells are formed. Figure 11.3 illustrates the stages of chromosome duplication and separation (*mitosis*) and cell division.

Figure 11.3
The stages of mitosis and cell division.

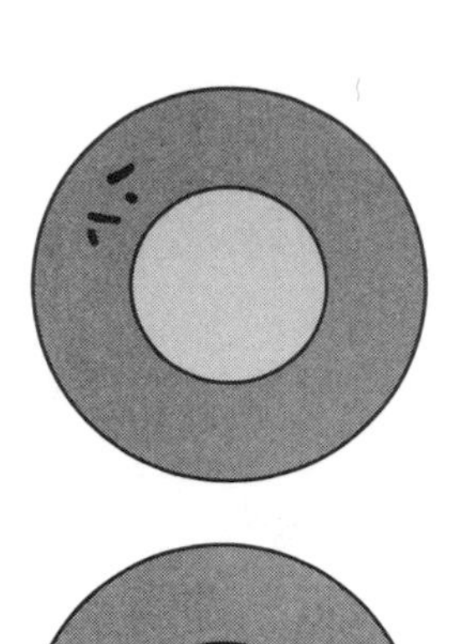

STAGE 1: a time of growth and metabolic activity. This is a period of time when the information in the DNA is being used to construct and organize every aspect of cellular life. During this time the DNA copies itself and by the end of this stage the cell contains two identical copies of its normal DNA content. At this stage no chromosomes are visible within the cell nucleus.

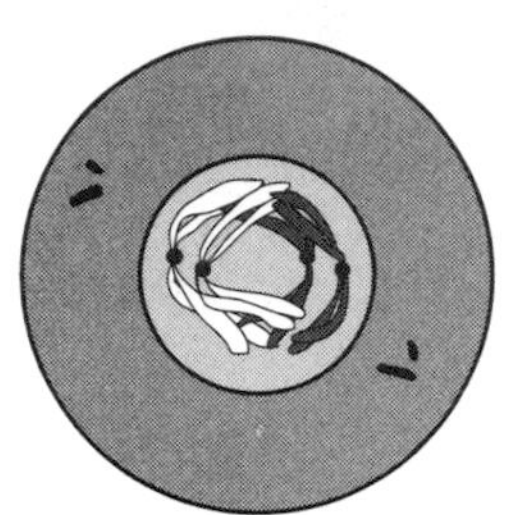

STAGE 2: the chromosomes become visible. In each chromosome the DNA coils around a protein core, and the duplicated copies are connected at a single site.

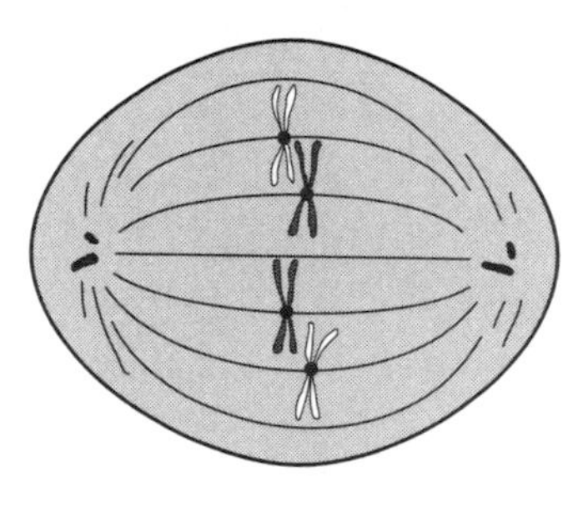

STAGE 3: the dissolving of the nuclear membrane. Protein "wires" or fibers form around the chromosomes which aid in the proper distribution of the chromosomes to the resulting daughter cells. The cell makes these fibers across itself from end to end, encasing the chromosomes in a scaffold-like structure. This structure is essential in moving the correct number of chromosomes to the proper position on either side of the cell.

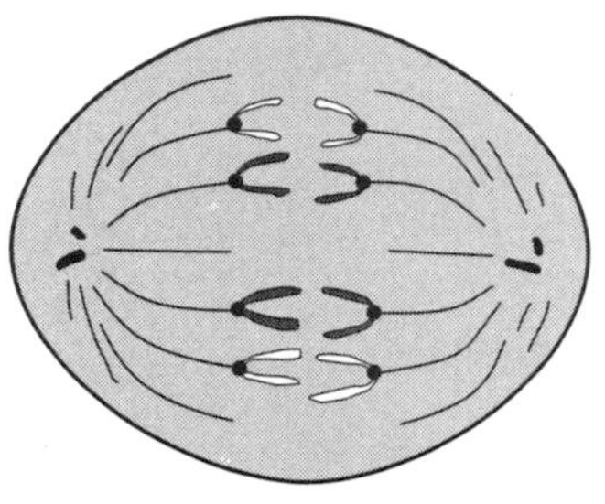

STAGE 4: the chromosome pairs actually separate and move to opposite sides of the cell. The fibers synthesized in Stage 3 are like tow lines that literally "haul" the chromosomes into position in an energy requiring process.

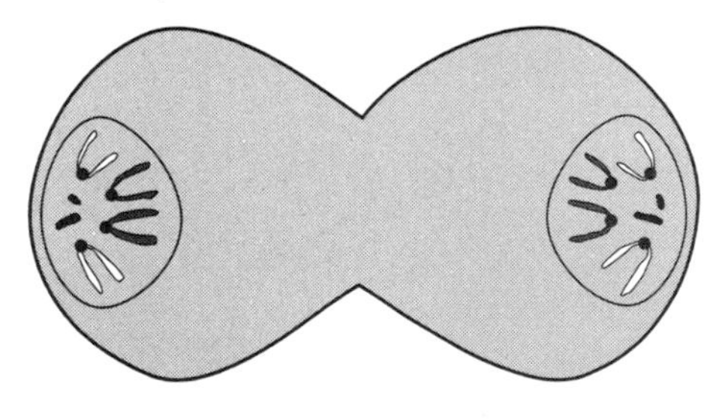

STAGE 5: the disassembly of the scaffolding, as the chromosomes are now encased in newly synthesized nuclear membrane. Once the two nuclei are formed the division of the cytoplasm begins. The cell membrane that surrounds the parent cell pinches inward. The two halves of the cell each contain organelles of the parent cell, ensuring "start-up operating" machinery for the essential metabolic processes of the daughter cells.

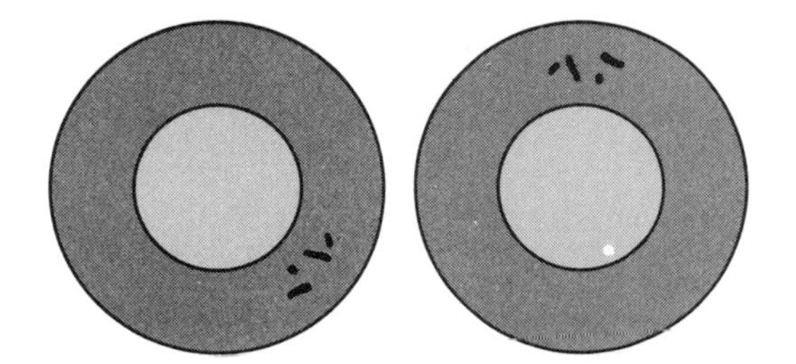

STAGE 6: the cell membrane completely separates the cytoplasm, resulting in two independent daughter cells. No chromosomes are visible at this stage.

Cell division in prokaryotic organisms is much less complex. The DNA of these organisms is not confined to a nucleus but floats, web-like, throughout much of the cell and is attached at various points to the cell membrane. When the cell has grown to a size where it must divide, DNA replicates. Using the sites of attachment as anchors the replicated DNA moves apart from the original DNA and a small fissure begins to grow inward from the cell membrane. New membrane and cell walls form between the DNA copies and eventually split the cell into two cells, each containing the required DNA information and necessary cytoplasmic machinery (see Figure 11.4).

Cell division in prokaryotic organisms is also the process for reproduction of the organisms. Some eukaryotic organisms also reproduce themselves by mitosis and cell division.

Repair and maintenance of a healthy organism relies on the ability of cells to divide in order to replace worn-out or damaged structures. When overexposure to the sun causes skin damage, for example, the injured and dead skin cells are sloughed off and cell division soon replaces the lost cells. Similarly, cell division is essential in wound healing. In a human adult, cell division replaces approximately 200 million worn-out cells per minute.

Although cell division is an essential process for growth, repair, and reproduction in many types of cells, in humans other types of cells, such as muscle cells and nerve cells, generally never divide after the embryonic stage of life. The number of cells remains fixed. If muscle or nerve cells are damaged they cannot be replaced. Muscle cells can grow by increasing the size of the cell itself. This can be seen in body builders whose muscle size will increase not as the result of an increase in the number of muscle cells but rather because of an increase in the size of individual muscle cells.

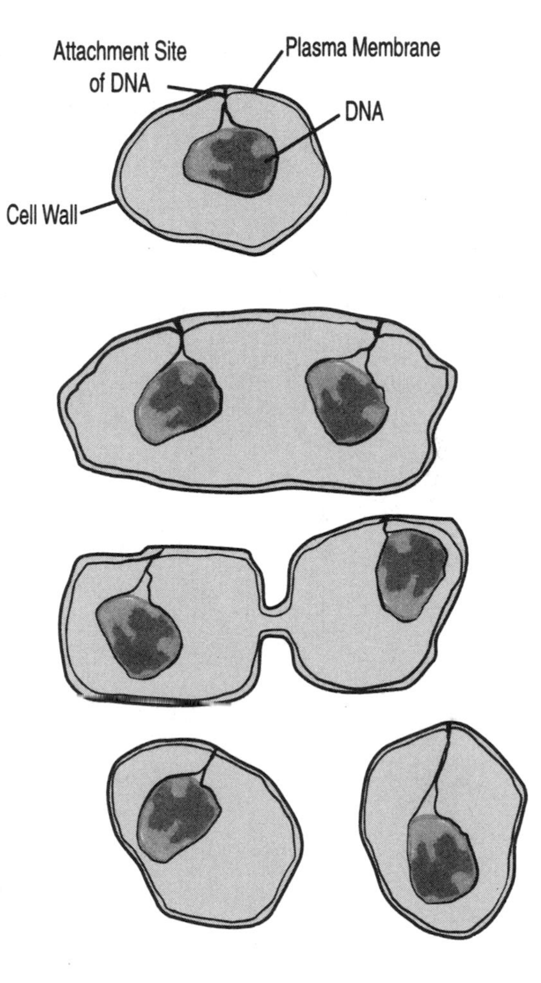

Figure 11.4
Bacterial cell division

ANALYSIS

Write responses to the following questions in your notebook:

1. Describe what might happen to a cell that grew very large but was unable to divide.
2. Why is it important that DNA copies are distributed equally during cell division?

3. Why is it important that each daughter cell receive cellular organelles such as mitochondria, chloroplasts, golgi apparatus, endoplasmic reticulum? Describe which process(es) a cell might be unable to accomplish if it were lacking one of these organelles.
4. Describe the stages of cell division that would require energy. Which stages would require that new biomolecules and cellular structures be made?
5. What difficulties might arise for an organism if muscle cells or nerve cells are damaged?

EXTENDING IDEAS

- Cancer is a disease that has a wide variety of symptoms and can affect one or more organs in the body. One common feature of all cancers is the loss of control over cell division. Research a specific form of cancer, describe what is known about changes that have occurred at the cellular level, and how the treatment reflects what is understood about how these changes cause the disease.
- Extensive research is being conducted to develop ways to stimulate nerve cells to divide in the hopes of helping spinal cord injury patients regain mobility in paralyzed limbs. Research why scientists believe that they can stimulate normally non-dividing cells to divide and how this may help these patients.

ON THE JOB

CELL BIOLOGISTS Cell biologists are scientists who study the cells of plants and animals. A cell biologist observes the growth and division of cells and conducts research to identify how chemical and physical factors might influence normal and abnormal cells. Some of this research focuses on identifying what might cause a cell to change from a normal growth pattern to an abnormal growth pattern. Cytologists apply the tools and techniques of cell biology to medicine in a clinical setting such as a hospital. Using a microscope, they examine cell samples to identify abnormalities in cell color, shape, or size that might be an early indication of cancer or another disease. Some cytologists are technicians who prepare and stain cell samples and might specialize in examining one particular type of cell. These technicians have specific training and receive certification in cytotechnology. The laboratory techniques and procedures used by cell biologists are not unique to cell biology, but are the techniques and tools used by chemists, geneticists, molecular biologists, or physiologists. With a four year college degree, positions as laboratory assistants or research cell biologists are possible. With a master's or doctoral degree cell biologists can teach in a university or pursue independent research in a university, or a laboratory. Classes such as biology, chemistry, math, English, and computer science are necessary.

Return to Reap the Spoils

Reaping the Spoils

INTRODUCTION This is your final analysis for the extended investigation on milk as an environment. Read the following Task section. You will need to write a final laboratory report and use the report for preparing a presentation.

TASK

1. With your group, analyze your data.
2. Discuss results and draw conclusions from the results. (Take notes on the discussion for use in writing your laboratory report.)
3. Plan your presentation. The presentation should include what observations and measurements your group chose to collect; the reasons your group chose to make those observations or collect the data; a summary of your results; and any conclusions your group was able to make based on your analysis. Include in your planning individual responsibility for preparing any illustrations, graphs, and text to be used in the presentation.
4. Write your own individual report, including the following information:
   - the question or questions being asked
   - the design of the experiment
   - the procedures used to collect data
   - the data
   - the analysis and significance of the data
   - conclusions based on the data and your understanding of resource use and metabolic processes of living organisms

- further questions you might wish to pursue and how you might go about answering them
- sources of possible errors

5. Use material from your report to prepare your portion of the presentation.

EXTENDING IDEAS

After the discovery of the existence of microorganisms by Anton van Leeuwenhoek in the seventeenth century, scientists began to investigate the origins of these life forms. Some believed that the microorganisms were formed spontaneously from the nonliving matter present in their environment. Others believed that the "seeds" or "germs" of these microscopic creatures were always present in the air and in foods and could grow, provided that the conditions were suitable for their development. This argument over the theory of spontaneous generation raged for years. Scientists gathered some evidence that the latter theory was more likely, but it wasn't until Louis Pasteur, in 1861, carried out a set of definitive experiments to prove that life only came from life that the spontaneous generation theory was dispelled.

Pasteur boiled a broth and placed it in a flask connected to the outside air only by a piece of tubing bent in such a way that microorganisms could not pass through. After many weeks, this broth failed to show any microbial growth. If the neck of such a flask was broken off so that air entered, the broth rapidly became populated with microbes.

State what you think Pasteur's conclusions were, based on your understanding of the milk experiment.

ON THE JOB

MEDICAL TECHNOLOGISTS Do you think you might be interested in applying your skills in performing laboratory techniques to the medical field? Medical technologists are professionals in the health care field who perform laboratory tests for detecting, diagnosing, and ultimately determining the treatment of diseases. A medical technologist may be responsible for interpreting the results of tests, reporting the findings to the physician, comparing test results with clinical data, or even recommending specific tests or sequences of tests to diagnose disease. A technologist working in a laboratory might operate sophisticated medical instruments and machines or conduct tests such as blood counts, urinalysis, and skin tests, or use a microscope to examine body fluids and tissue samples. Technologists with advanced degrees may choose to specialize in a specific field (such as cytology, chemistry, or biochemistry) or in managerial or supervisory roles (such as managing sections of a laboratory or supervising lab technicians). Medical technologists are employed in a variety of settings such as research, education, veterinary science, public health, or with a company designing new diagnostic equipment. Medical technologists usually have a four year college degree and have completed a program in medical technology. After completing these studies, technologists can apply for certification by the professional organization of medical technologists. Classes such as biology, chemistry, physics, English, and computer science are necessary.

Where Will It All End?

PROLOGUE In previous learning experiences, you have focused on the characteristics of life and the resources and processes required to maintain life. A critical question that seems to follow naturally from this study of "life" is: What happens when these resources are not accessible and these processes no longer occur? For example, what if food were no longer available? The characteristics of life that depend on the uptake and transformation of the nutrients into materials and energy for the organism would disappear. Unless the organism could find a way to locate and get food, that organism would perish. Death marks the end of the complex chemical reactions and transfer of energy that enable organisms to grow, respond, reproduce and maintain themselves. Is death an end to life or an integral part of the cycle of life?

When Organisms Die

When organisms die, the process of decomposition returns the biomolecules which had made up the living organism to the soil and air. Microorganisms called decomposers break down the proteins, fats, carbohydrates, and other complex organic substances and transform them into smaller molecules of carbon dioxide, ammonia, and other simple inorganic compounds. Thus, decomposers both rid the earth of no longer living organic debris, and return to the air and soil the simple compounds that plants require for the synthesis of their food and biomolecules. Life then goes full circle. Using carbon dioxide from the air, minerals from the soil, and energy from the sun, plants generate the food that they and other organisms require to synthesize the biomolecules they use to maintain the characteristics of life. When these organisms die, they return these simple molecules to the soil and air for reuse. If it were not for that cycle, the world would be filled with fossilized remains of once-living plants and animals.

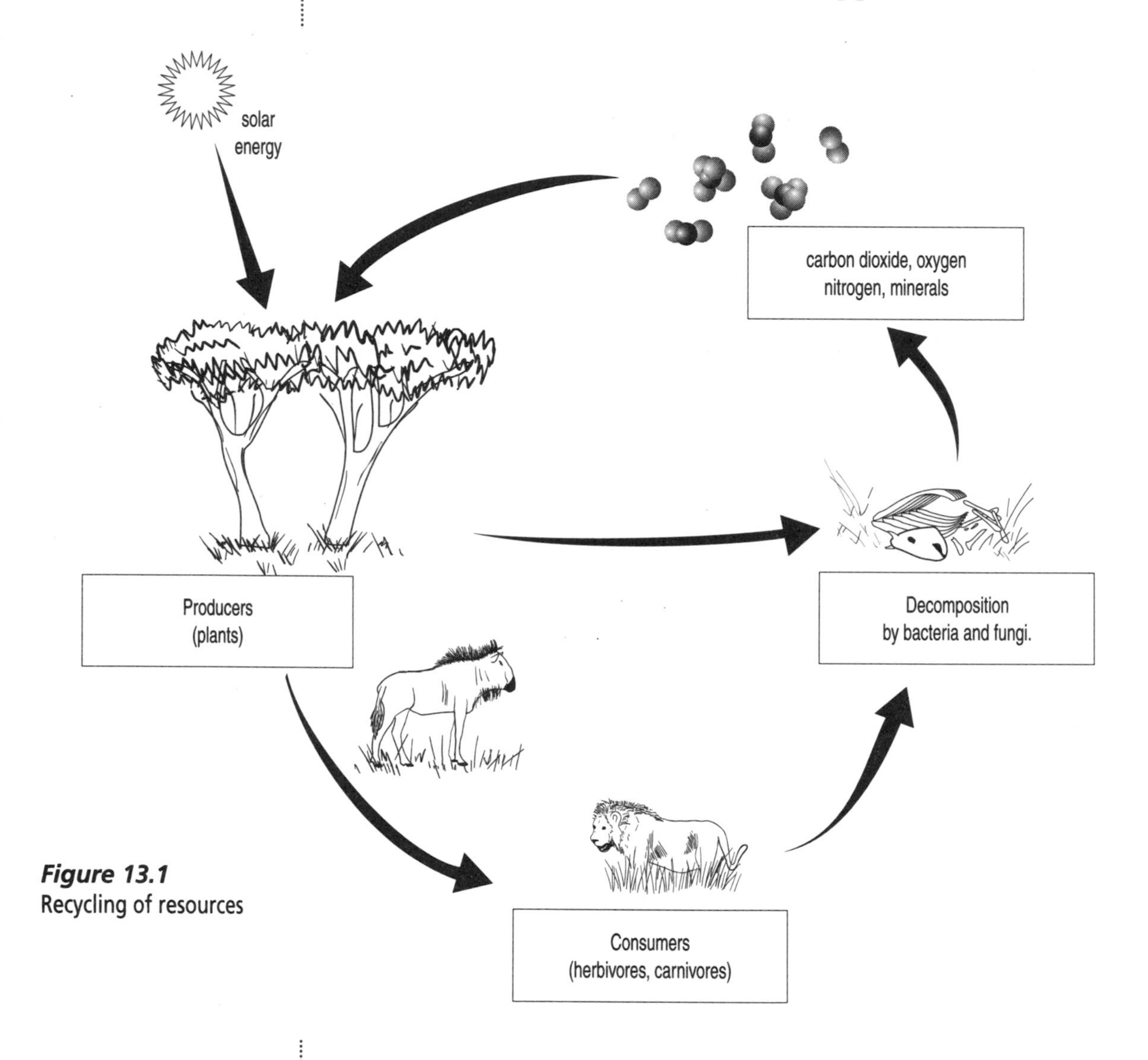

Figure 13.1
Recycling of resources

ANALYSIS

Examine the diagram (Figure 13.1) and write responses to the following questions:

1. For millions of years plants and animals have been living on the earth and using resources. Explain why the earth has not exhausted its supply of nutrients, minerals, carbon dioxide, water, and oxygen.

Include information from the diagram, from the reading "When Organisms Die," and from other material and your experience with this module.

2. What do you think might happen if living things did not die?
3. Explain the following statement: The atoms of carbon in your body may have been part of a *Tyrannosaurus rex*; the oxygen you breathe may have been breathed by Julius Caesar.

NEVER SAY DIE

INTRODUCTION If there were an endless supply of resources, could an organism live forever? Is death inevitable? Is death the only outcome of life? Is there such a state as immortality? Is immortality a desirable state?

Science fiction is filled with stories of quests for immortality and the consequences of achieving it. What would it be like to live forever or for at least 500 or 600 years? What would it be like if everyone in the world could live forever? What might you look like? How would you plan your life? How might the quality of life change? What would the world be like?

TASK

Imagine that the year is 2120. You and many other peoplc have decided to undergo a treatment which might allow you to live forever, or at least for a very long time. No one is sure how long you will survive after the treatment since it only recently has been made available to humans; however, prolonged testing on mice has indicated that human life expectancy could be extended for at least 500 years.

Write a short science fiction story describing what your life will be like in the coming 500 years. Be sure to include personal, biological (characteristics to maintain life), social (economic and political), environmental, and ethical issues as well as consequences that may result from people living forever.

EXTENDING IDEAS

- A recent theory in developmental biology suggests that the lives of humans and other multicellular organisms depend in a variety of ways upon the programmed death of our own cells. Through a process called *apoptosis*, healthy cells prepare the means for their own destruction, then commit suicide. In development, this healthy suicide is the process behind the loss of the tail of the tadpole as it turns into a frog; it is responsible for the loss of the juvenile tissues of the caterpillar when the time comes for metamorphosis into a butterfly. In human beings, apoptosis deletes cells from the embryo's developing limbs, leaving behind ten little fingers and toes. In our immune system, apoptosis plays a number of crucial roles: dying white blood cells self-destruct so that they can be cleaned up by other cells in the body's defense system. Cells infected by viruses may actually self-destruct so that the body's immune system can clean up the mess. Today, researchers wonder if apoptosis may hold the key to stopping certain cancers: if cancer represents uncontrolled cell growth, could the growth be halted by triggering programmed cell death? Does the discovery of apoptosis affect our understanding of what life itself means?

Are there any organisms that do not die? In an article entitled "The Immortal Microorganism" (*New Scientist*, May 20, 1989) microbiologist John Postgate states "When [bacterial] cell division takes place, there is no way of telling which of the two short rods is the mother and which is the daughter. Indeed, such evidence as is available tells us that no mother-daughter relationship exists; both progeny are of equal age.... Did the parent cell grow old and die? Of course not. So, do klebsiellae [a kind of bacterium] have any equivalent of aging and death, analogous to these processes in, for example, ourselves?" How would you answer that question?

ON THE JOB

BIOLOGICAL TECHNICIAN Are you interested in using your organizational and detail-oriented skills to work in a variety of settings? Biological technicians assist biologists as they study living organisms. The work of a biological technician may be as diverse as pharmaceutical or medical research, biotechnology research and development, or in forensics (in police crime laboratories). Workers in pharmaceutical and medical research might help find cures for AIDS, cancer, or other diseases. Biotechnology research and development might result in the development of new products such as food additives, insecticides, or fertilizers. When genetic engineering is a focus of biotechnology research, the technician's role may be to aid in transferring genetic information from one kind of plant or animal to another. This can result in crops having better yields or being resistant to low temperatures. In a police crime lab, the work might be analyzing criminal evidence. Biological technicians perform lab experiments using equipment such as microscopes, chemical scales, or centrifuges, control the conditions of the experiment, record instrument readings, or analyze data to identify trends and patterns. Depending on the nature of the research, biological technicians might work with lab animals such as mice, rats, or guinea pigs. With a high school diploma, one year of technical training is necessary for a position as a biological technician. With additional training, technicians can advance to positions as higher level technicians or supervisors. Classes such as biology, chemistry, physiology, math, and English are recommended.

ECOLOGIST Are you interested in how both the natural world and the man-made world can coexist? Ecologists study the relationships between organisms and their environment and the factors that influence these relationships. Ecologists may focus on studying organisms and their habitats, fluctuations in population size, or ways of preserving environments. Ecologists may do basic research to study how the environment works, then apply this knowledge by working with other professionals to preserve the environment, or propose solutions to fix damage already done. Ecologists work with lawyers, economists, and policy makers to try and balance the need to use resources in the environment in order to sustain life with the need to preserve habitats and species. Ecologists try to understand the implications of environmental problems and to propose practical solutions. They research and write environmental impact studies before construction of roads or buildings take place, work on new systems to restore water quality to streams and rivers or to purify waste from sewer systems. Ecologists were involved in the proposal to flood the Grand Canyon in early 1996 in order to restore the natural ecosystem of the canyon. In the field, an ecologist would collect data by conducting experiments and making careful observations. Scientific and mathematical models are used in conjunction with field work. Four year college degrees are needed for nonresearch positions such as testing and inspection. With a master's degree applied research, inspection, or managerial positions are available. With a doctorate, teaching in a university and doing independent research are possible. Classes such as biology, chemistry, animal physiology, chemistry, geology, and English are recommended .

Suspended Between Life and Death

PROLOGUE In the late twentieth century, the ethics, the economics, and the humanity of extending a human life *when the quality of that life has been compromised* has become a pressing issue. Today, huge numbers of Alzheimer's patients survive long after they have lost the capacity to think for themselves. Victims of total paralysis can be sustained through the application of respirators and feeding tubes. Modern medicine is often able to extend lives at the threshold of death.

Modern technology is forcing us to think again about our understanding of life and death; Just as old definitions of life have been challenged by new discoveries in science, so old definitions of death no longer apply. For example, in an 1890 law the U.S. courts defined death as the cessation of the circulation of blood, of the pulse, and of respiration. But as we all know, in the modern world of emergency medicine many thousands of Americans whose respiration and/or circulation has stopped have been restored and then returned to productive, healthy lives.

One group of patients in which the number of compromised survivors has increased dramatically is those whose heartbeats or respiration have been restored after their brains have already been damaged. Today, in the U.S., there are between 5,000 and 10,000 brain-damaged patients kept alive by artificial life support but are unconscious and, for the most part, have no hope of full recovery. The nation's medical bill to sustain comatose and vegetative patients (the two diagnoses are somewhat different) amounts to approximately $1 billion per year. It is these types of patients who will be the focus of the final learning experience in this module.

The Problem Defined

INTRODUCTION Thus far in this module, you have constructed a scientific definition of life. One facet of that definition is that all living things must die. There are certain life processes without which life cannot be sustained. The question arises, then, is a person alive if he is unable to sustain his life processes on his own but medical technology is able to perform those processes for him?

What remains is to reflect upon the definition of *human* life and the quality of life. If the technical question of what makes a living thing *alive* is a topic for biologists, the question of what makes a human life worth living is more a question of philosophy, of ethics, and of religion. Perhaps there is no better place to clarify our views about life than around the question of death, because it urges the living to reflect upon the experience of life.

To reflect about what constitutes human life, you are going to simulate an academic conference on the bioethics of life and death. Each of you will be given a role to play at the conference.

TASK

For your role, **prior to the conference**, you should do all of the following:

- Read the fictional news article, "A Difficult Choice" as background for the conference. Also, read the "role profile" notes which your teacher gives you. Keep your "role profile" secret. Nobody else should know who you are or what your point of view will be prior to the conference. If a little suspense and mystery are maintained before the conference, it will be more exciting for everyone involved.
- Enter into the role you have been assigned. **From the perspective of that role**, write a thoughtful two- to three-minute opening statement, to be read or spoken aloud during the conference. Treat the information provided to you about your role as only an *outline* for the character you will create. Go beyond what is given to you on the Role Description sheet, expand upon the rationales for your opinions, and develop a "personality" for the role. Take some time to consider what your character would think about issues not covered in your character outline. For example: your role description states that your character is interested in saving money. By extension, what do you think such a character would think about using heroic measures to save a life? It is important that you give some thought

to the general world view of your character. It is very important that your character's viewpoint remains logical and consistent throughout the simulation.

- Bring your notes to the simulation, **as they will be collected after the conference ends**.

During the conference:

- You will be evaluated not only for your opening remarks but also for your thoughtful attention to other conference attendees' ideas. It is expected that you will rebut those who disagree with your character's point of view; it is likewise expected that you will ask questions and challenge the other speakers when you think doing so will broaden or deepen the conference discussion.

Following the conference:

- You will write a policy position statement. In this essay, you will be drawing from all that was said at the conference and from your own personal perspective to explain your personal views about the question discussed at the conference. The detailed requirements for this assignment are provided in the activity description "Is Life Extended?" at the beginning of the simulation.

Rules for conference debate:

- Your teacher is the conference moderator. Please respect the moderator's role.
- Raise your hand when you wish to speak. If the moderator calls for order in the conference, please respect this request.
- Fill out a name card with your character's name, and place this card in front of your chair at the conference. During the conference, please be sure to address other conference participants by their formal titles.
- Please feel free to raise questions in response to other participants' opening remarks, but also respect the moderator's responsibility for time management. It is important that all conference attendees have a chance to make their opening remarks; in order to ensure this, it may be necessary for the moderator to limit questions and debate at some points.
- During the simulation, your performance may be evaluated by visiting guests.

A Difficult Choice

January 20, 1999

CITY NEWS—This week, following months of unusually well-publicized preparation, the City University Department of Bioethics will convene a conference to discuss the ethics of long-term life support. For a variety of reasons, this conference is receiving more than its share of national press attention. Several senators prominent in the resurfacing national debate over health care are expected to attend. For weeks there have been rumors that key members of Congress are considering new legislation on the right-to-die issue; speculation centers on whether one or another senator may use this conference as an opportunity to propose new legislation. Expected to attend are a wide range of health care professionals, several legal scholars, and the family members of several comatose patients. All of these people hope to influence any legislation which may make it to a vote in Congress.

In addition, the focus of discussions at the day-long conference is the fate of Mark Vorst. If the celebrity status of many of the conference's attendees weren't enough to draw a media blitz, the conference's focus on Mr. Vorst's tragic story would itself be certain to turn many heads. By now, the story of Mark Vorst is familiar to many Americans. Mr. Vorst, a rising star in the Senate and a leading advocate of health care reform back in 1994, suffered severe brain damage when his car careened off a bridge during an ice storm. Since that time, Vorst has been kept alive by surgically implanted feeding tubes and a respirator. Medical experts agree that he is in a "persistent vegetative state," with no chance of regaining consciousness; however, his face does occasionally register a smile or a grimace, and from time to time his eyes produce tears.

Because the health care advocate himself had neglected to fill out either a health care proxy or a living will, Mr. Vorst's own wishes in the event of such a catastrophe are not known. As a result, his fate has become the subject of a bitter five year court battle. Mr. Vorst's parents have fought for the right to maintain their son on life support, while his insurer and his wife have both been fighting for the right to remove the support. This case is almost a mirror image of the Nancy Cruzan case, settled by the Supreme Court in 1990. In that case, over the protests of Cruzan's doctors, the family won the right to stop life support.

To date, the life-support care for Mr. Vorst has cost the U.S. taxpayers approximately $650,000, which is on top of the $200,000 in emergency and intensive care treatment Mr. Vorst received at the time of his injury (as a member of the U.S. Senate, Mr. Vorst's health insurance was provided by the federal government). The court battle, now scheduled for a spring hearing before the Supreme Court, has inspired a nationwide debate about the role of medicine and the question of limit-setting in health care spending. The case may never receive a hearing before the Supreme Court, if Congress chooses to legislate the issue before the Court's spring session.

At the conference, participants will be asked to join in a dialogue about the same questions which the Supreme Court or Congress may soon debate:

"Should Mark Vorst be left on life support? Should there be national legislation setting limits to medical care? If so, what should those limits be? What are the ethical, economic, and social implications of our answers to these questions?"

Is Life Extended?

TASK

The policy-position essay you are to write is composed of a number of parts. Be sure you incorporate all parts of the question in your essay.

- Based on what you have learned from this module, do you feel that a brain-damaged person on life-support machines is alive? Use the biology from this module to defend your answer. Describe where biology may or may not be sufficient in answering this question.
- Should a brain-damaged, unconscious patient be left on life support? If so, for how long? At what cost? Should there be limits to such life-support care? Why or why not?
- Should there be national legislation limiting medical care for people whose consciousness has been compromised? If so, what should those limits be, and why? If not, why not?
- What are the societal implications of your answers to these questions? What do your answers suggest about how our society should invest its resources?
- How has your response to the above questions been influenced by your family's attitudes? By your friends? By your experiences? By the American culture around you?
- Conclude by reflecting upon how this simulated experience has deepened your thinking about what, if anything, makes human life unique.

Your essay will be evaluated on some or all of the following criteria:

- completeness (have you addressed all six parts of the essay question?);
- organization of ideas;
- reasoned basis for your responses;
- thought-provoking conclusion; and
- clarity of writing.

EXTENDING IDEAS

- Research one of Dr. Jack Kevorkian's assisted suicide cases and make some comparisons between that case and the one discussed in the simulation.

ON THE JOB

LAWYER Do you think you would be interested in ensuring that an individual's rights are not violated? Lawyers or attorneys are trained to represent either the state or a client in the laws that govern society. It is their job to help clients understand their rights under the law and then to attain these rights in a court of law or government agency. Lawyers may work for businesses, advising on any legal matter, arranging for stock to be issued, consulting on tax matters, or real estate dealings. They may also work for individuals, advising clients on buying or selling a home, drawing up wills, or acting as a trustee or guardian. The training lawyers receive and the subsequent licensing allows them to practice law in any area, although many lawyers specialize in one type of law. Some lawyers don't work in courtrooms at all, and not all of any lawyer's time is spent in a courtroom. Many hours of preparation work takes place before a trial. Civil lawyers practice private law which deals with violations of contracts, deeds, leases, and mortgages. Probate lawyers specialize in planning and settling estates, and drawing up wills or deeds of trust. Corporate lawyers advise companies on their legal rights, obligations and privileges. Criminal lawyers focus on cases where an offense against society, such as theft or murder, has been committed. A background in science is useful for lawyers to work with scientific organizations such as biotechnology companies, or if they need to understand genetic evidence for a criminal trial, or to write living wills. Some lawyers may hold positions as researchers or professors in law school, and some people use their training in law in business positions. A high school degree, four year college degree and three years of law school are needed to become a lawyer.

Before being allowed to practice in a state, a lawyer must pass the bar exams of that state. Classes in biology, English, and history are recommended.

NURSE Nurses are health care professionals who work with individuals and groups of people to promote health, prevent disease, and care for the sick or injured. Nurses might work in a hospital or in another health care facility such as a physician's office, at a public health agency, school, camp or for a Health Maintenance Organization (HMO). Nurses work in collaboration with physicians and other members of a health care team to assess and monitor the condition of a patient and then to develop and implement a plan of health care. In a hospital, nurses work in many hospital units (such as the emergency room, maternity ward, or intensive care unit), and the responsibilities may vary from unit to unit. A general duty nurse might take patients' temperatures, pulse or blood pressure, administer medication or injections, record symptoms and progress of patients, change dressings, and assist patients with personal care, as well as other duties. Surgical nurses are responsible for preparing operating rooms, sterilizing instruments, and assisting surgeons during operations. Maternity nurses help in the delivery room, take care of newborns, and teach mothers how to feed and care for their babies. With additional training to expand skills and knowledge, registered nurses may become nurse practitioners and do many tasks previously handled by physicians (such as making diagnoses and recommending medications). Or with a master's degree registered nurses may become clinical nursing specialists and focus in one field of nursing (for example, working with cancer patients or in a cardiovascular unit). There are several different program sequences that provide nursing training. With a two year program in nursing, it is possible to work in any of the settings described above. With a four year college degree in nursing it is possible to hold an administrative position or work for a public health agency. With a master's degree it is possible to specialize or to teach in a school of nursing. In all cases, nurses must pass a licensing exam before they are registered nurses and can practice nursing. Classes in subjects such as biology, chemistry, math, physics, and English are necessary.

Nutrient Content of Food

The following table is a list of calorie, protein, fat and carbohydrate values by common serving sizes. All calorie values have been rounded to the nearest five calories. Values for the remaining nutrients are rounded to the nearest gram. The abbreviation (Tr) means there is a trace amount of that nutrient present in the quantity described. This table has been abstracted from *Nutritive Value of Foods* (Home and Garden Bulletin Number 72) available from the Superintendent of Documents, U.S. Government Printing Office, Washington, D.C. 20402.

FOOD	AMOUNT	CAL-ORIES	PRO-TEIN	FAT	CARBO-HYDRATE
Almonds, unsalted	1 oz.	165	6	15	6
Apple	1 medium (2 3/4-inch diameter)	80	0	0	21
Apple Juice (Cider)	1 cup	115	0	0	29
Applesauce:					
sweetened	1 cup	195	0	0	51
unsweetened	1 cup	105	0	0	28
Apricots:					
dried	1 cup (28 halves)	310	5	1	80
fresh	3 medium	50	1	0	12
Asparagus, raw/cooked	4 spears	15	2	0	3
Avocado	1 8-ounce avocado	305	4	30	12
Bacon:					
regular	3 slices	110	6	9	0
Canadian	2 slices	85	11	4	1
Banana	1 medium	105	1	1	27
Bagel	1(3 1/2 inch diameter)	200	7	2	38
Barley, pearl	1 cup, raw	700	16	2	158
Beans:					
green, fresh	1 cup	30	2	0	7
kidney, canned	1 cup	230	15	1	42
snap, canned	1 cup	45	2	1	10
Beef:					
cooked, lean meat only	2.8 ounces	175	25	8	0

FOOD	AMOUNT	CAL-ORIES	PRO-TEIN	FAT	CARBO-HYDRATE
flank, uncooked, lean meat only	4 ounces	165	24	65	0
porterhouse, uncooked, lean meat only	4 ounces	185	24	9	0
rib roast, cooked, lean meat only	2.2 ounces	150	17	9	0
round, cooked, lean meat only	2.6 ounces	135	22	5	0
rump, uncooked, lean meat only	4 ounces	180	24	9	0
sirloin, broiled, lean meat only	2.5 ounces	150	22	6	0
uncooked, as purchased, no bone	1 pound	1,510	74	132	0
Beef, dried	2.5 ounces	145	24	4	0
Beef Stew	1 serving (1 cup)	220	16	11	15
Beets	2 (1/2 cup, diced)	30	1	0	7
Biscuits w/baking powder & enriched flour	1 (2-inch diameter)	100	2	5	13
Blackberries, fresh	1 cup	75	1	1	18
Blueberries, fresh	1 cup	80	1	1	20
Bologna	2 slices	180	7	16	2
Bouillon cubes (all types)	1 packet	15	1	1	1
Brazil Nuts	1 ounce	185	4	19	4
Breads, fresh or toasted:					
Boston brown	1 slice	95	2	1	21
cracked wheat	1 slice	65	2	1	12
French	1 slice	100	3	1	18
Italian	1 slice	85	3	0	17
rye, light	1 slice	65	2	1	12
white, enriched	1 slice	65	2	1	12
raisin	1 slice	65	2	1	13
whole wheat	1 slice	70	3	1	13
Bread Crumbs, soft	1 cup	120	4	2	22
Broccoli, fresh cooked	1 cup	45	5	0	9
Brussels Sprouts	1 cup	60	4	1	13
Butter	1 tablespoon	100	0	11	0
Buttermilk	1 cup (8 ounces)	100	8	2	12
Cabbage:					
cooked	1 cup	30	1	0	7
raw	1 cup shredded	15	1	0	4
Cakes:					
angel-food	1/12 of 10-inch cake	125	3	0	29
cheesecake	1/12 of 10-inch cake	280	5	18	26
devil's food cake w/ choc. frosting	1/16 of 9-inch cake	235	3	8	40
cupcake, frosted	1 medium	120	2	4	20
fruitcake	1/32 of 8-inch loaf	165	2	7	25
poundcake	1/17 of 8-inch cake	120	2	5	15
yellow cake, choc. frosting	1/ 16 of 9-inch cake	235	3	8	40
Candy:					
caramel, plain	1ounce	115	1	3	22
chocolate, milk	1 ounce	145	2	9	16
fudge	1 ounce	115	1	3	21
gum drops	1 ounce	100	0	0	25
hard candy	1 ounce	110	0	0	28

FOOD	AMOUNT	CAL-ORIES	PRO-TEIN	FAT	CARBO-HYDRATE
Cantaloupe	1/2 (5-inch diameter)	95	2	1	22
Carrots:					
fresh cooked	1 cup, sliced	70	2	0	16
raw	1 cup	45	1	0	11
	1whole	30	1	0	7
Cauliflower					
cooked	1 cup	30	2	0	6
raw	1 cup	25	2	0	5
Celery, raw	1 cup, diced	20	1	0	4
	1 stalk	5	0	0	1
Cereals, cooked:					
corn grits, enriched	1 cup	145	3	0	31
farina, enriched	1 cup	105	3	0	22
rolled oats, oatmeal	1 cup	145	6	2	25
Cereals, ready-to-eat:					
All Bran	1/3 cup (1 ounce)	70	4	1	21
Corn Flakes	1 1/4 cup (1 ounce)	110	2	0	24
Rice Krispies	1 cup (1 ounce)	110	2	0	25
Shredded Wheat	2/3 cup (1 ounce)	100	3	1	23
Cheese:					
American, processed or natural	1 slice (1 ounce)	105	6	9	0
Camembert	1 wedge	115	8	9	0
Cheddar	1 ounce	115	7	9	0
cheese spread	1 ounce	80	5	6	2
cottage, creamed	1 cup	215	26	9	6
uncreamed	1 cup	125	25	1	3
cream	1 ounce	100	2	10	1
Mozzarella, whole milk	1 ounce	80	6	6	1
Parmesan	1 tablespoon	25	2	2	0
Ricotta, partially					
skimmed milk	1 cup	340	28	19	13
whole milk	1 cup	430	28	32	7
Roquefort or blue	1 ounce	100	6	8	1
Swiss	1 slice (1 ounce)	105	8	8	1
Cherries					
canned, red sour, pitted	1 cup	90	2	0	22
fresh, sweet	10 cherries	50	1	1	11
Chicken, fried w/skin					
breast	4.9 ounces	365	35	18	13
drumstick	2.5 ounces	195	16	11	6
roasted, flesh only					
breast	3 ounce	140	27	3	0
drumstick	1.6 ounce	75	12	2	0
Chicken Pot Pie	1 3-inch pie	545	23	31	42
Chili Con Came, canned	1 cup	340	19	16	31
Chocolate:					
milk	1 ounce	145	2	9	16
semi-sweet	1 ounce	145	1	10	16
unsweetened (baking)	1 ounce	145	.3	15	8
Chocolate Milk Drink 1%	1 cup	160	8	3	26
Chocolate Syrup	2 tablespoons	85	1	0	22
Chop Suey	1 cup	300	26	17	13
Clams, uncooked	3 ounces, meat only)	65	11	1	2
canned	3 ounces	85	13	2	2
Cocoa powder	3/4 ounces	75	1	1	19
Cocoa, prepared with milk	1 serving	225	9	9	30

FOOD	AMOUNT	CAL-ORIES	PRO-TEIN	FAT	CARBO-HYDRATE
Coconut, fresh shredded	1 cup	285	3	27	12
Coffee, brewed	6 ounces	0	0	0	0
Cola Drinks	12-ounce bottle	160	0	0	41
Cookies:					
chocolate chip, commercial	4 cookies	180	2	9	28
fig bar	4 squares	210	2	4	42
oatmeal w/raisins	4 cookies	245	3	10	36
sandwich-type	4 cookies	195	2	8	29
sugar	4 cookies	235	2	12	31
Corn					
canned, whole kernel	1 cup	165	5	1	41
fresh	1 ear	85	3	1	19
Cornmeal, uncooked, enriched	1 cup	500	11	2	108
Crabmeat, canned	1 cup	135	23	3	1
Crackers:					
graham	2 crackers	60	1	1	11
saltine	4 crackers	50	1	1	9
rye wafers	2 wafers	55	1	1	10
Cranberries, raw	1 cup	45	0	1	11
Cranberry Juice Cocktail	l cup	145	0	0	38
Cranberry Sauce canned	1 cup	420	1	0	108
Cream:					
half-and-half	1 tablespoon	20	0	2	1
heavy	1 tablespoon	50	0	6	0
	1 cup	820	5	88	7
light	1 tablespoon	30	0	3	1
	1 cup	470	6	46	9
sour, commercial	1 tablespoon	25	0	3	1
Cucumbers, raw	6 slices (l/8-inch thick)	5	0	0	1
Custard, baked	1 cup	305	14	15	29
Dandelion Greens, cooked	1 cup	35	2	1	7
Danish Pastry, plain	4 1/4 inch piece	220	4	12	26
Dates, whole, pitted	10 dates	2302	0	61	
Doughuts (cake type)	1 doughnut	210	3	12	24
Eggs					
fried	1 egg, margarine	90	7	10	1
scrambled	1 egg, milk and margarine	100	7	7	1
shirred or poached	1 egg	75	6	5	1
Eggplant, steamed	1 cup	25	1	0	6
Figs, dried	10 figs	475	6	2	122
Fish Sticks	1 fish stick	70	6	3	4
Flounder, baked	4 ounces	80	17	1	0
Flour:					
all-purpose, enriched	1 cup, sifted	420	12	1	88
cake, unenriched	1 cup, sifted	350	7	1	76
whole wheat	1 cup, sifted	400	16	2	85
Frankfurters	1 frankfurter	145	5	13	1
French Toast	1 slice (no syrup)	155	6	7	17
Fruit Cocktail, canned	1 cup fruit and syrup	185	1	0	48
Gelatin Dessert	1/2 cup prepared	70	2	0	17
Gelatin, unflavored, dry	1 envelope	25	6	0	0
Ginger Ale	12 oz. bottle	125	0	0	32
Gingerbread	1/9 of 8-inch cake	175	2	4	32
Grapefruit	1/2 medium	40	1	0	10

FOOD	AMOUNT	CAL-ORIES	PRO-TEIN	FAT	CARBO-HYDRATE
Grapefruit Juice:					
sweetened, canned	1 cup	115	1	0	28
unsweetened	1 cup	95	1	0	22
Grapefruit Sections	1 cup with syrup	150	1	0	39
Grape Juice	1 cup	155	1	0	38
Grapes	10 grapes	35	0	0	9
Gravy:					
canned beef	1 cup	125	9	5	11
brown from mix	1 cup	80	3	2	14
Haddock, breaded,fried	1 fillet (3 ounces)	175	17	9	7
Halibut, broiled w/but.	3 ounces	140	20	6	0
Herring, pickled	3 ounces	190	17	13	0
Honey	1 tablespoon	65	0	0	17
Honeydew Melon	1/10 of 6 1/2' inch melon	45	1	0	12
Ice Cream, vanilla	1 cup (11 percent fat)	270	5	14	32
	1 cup (16 percent fat)	350	4	24	32
Ice Milk	1 cup (4% fat)	185	5	6	29
Jam	1 tablespoon	55	0	0	14
Jelly	1 tablespoon	50	0	0	10
Kale, cooked	1 cup	40	2	1	7
Ketchup	1 tablespoon	15	0	0	4
Lamb:					
chop, lean, braised	1.7 ounces	135	17	7	0
leg, lean, roasted	2.6 ounces	140	20	6	0
rib roast, lean	2 ounces	130	15	7	0
Lemon Juice	1 cup	60	1	0	21
Lettuce, iceberg	1 cup	5	1	0	1
	1 head	70	5	1	11
Liver	3 ounces	185	23	7	7
Macaroni and Cheese	1 cup	430	17	22	40
Macaroni, cooked, enriched	1 cup	155	5	1	32
Mangos	1 medium	135	1	1	35
Margarine					
80% fat	1 tablespoon	100	0	11	1
40% fat	1 tablespoon	50	0	5	0
Marshmallows	1 ounce	90	1	0	23
Mayonnaise	1 tablespoon	100	0	11	0
Milk:					
buttermilk, cultured	1 cup	100	8	2	12
condensed, sweetened	1 cup	980	24	27	166
evaporated whole	1 cup	340	17	19	25
liquid, skimmed	1 cup	85	8	0	12
liquid, 98% fat-free	1 cup	120	8	5	12
liquid, whole	1 cup	150	8	8	11
nonfat dry milk, instant (dry powder)	1 envelope	325	32	1	47
Yogurt (made with partially skimmed milk): plain	1 cup	145	12	4	16
fruit flavored	1 cup	230	10	2	43
Molasses, light	2 tablespoons	85	0	0	22
Muffins:					
blueberry	1 muffin	135	3	5	20
bran	1 muffin	125	3	6	19
corn	1 muffin	145	3	5	21
English	1 muffin	140	5	1	27

FOOD	AMOUNT	CAL-ORIES	PRO-TEIN	FAT	CARBO-HYDRATE
Mushrooms:					
canned	1 cup	35	3	0	8
fresh	1 cup	20	1	0	3
Noodles, egg, cooked, enriched	1 cup	200	7	2	37
Oils:					
corn, cottonseed, olive, peanut, and soybean	1 tablespoon	125	0	14	0
Olives:					
green	4 medium	15	0	2	0
ripe	3 small	15	0	2	0
Onion, raw	1 cup	40	1	0	8
Orange	1 medium (2 5/8-inch diameter)	60	1	0	15
Orange Juice, fresh, frozen or canned	1 cup	110	2	0	26
Orange Sections	1 cup	85	2	0	21
Oysters, raw	1 cup (meat only)	160	20	4	8
Pancakes, from mix	1 (4-inch diameter)	60	2	2	8
Parsley, raw	10 sprigs	5	0	0	1
Peaches:					
canned, syrup pack	1 cup	190	1	0	51
fresh, uncooked	1 medium	35	1	0	10
Peanuts, roasted	1 cup	840	39	71	27
	1 ounce	165	8	14	5
Peanut Butter	1 tablespoon	95	5	8	3
Pears:					
canned, syrup pack	1 cup	190	1	0	49
fresh, uncooked	1 medium	100	1	1	25
Peas:					
green, canned	1 cup	115	8	1	21
split, dried, cooked	1 cup	230	16	1	42
Pecans	1 ounce	190	2	19	5
	1 cup	720	8	73	20
Peppers, green, raw	1 medium	20	1	0	4
Perch (ocean), breaded, fried	1 filet	185	16	11	7
Pickles:					
dill	1 medium	5	0	0	1
sweet	1 small	20	0	0	5
Pie Crust:					
homemade, double crust	1 9-inch shell	1,800	22	120	158
homemade, single crust	1 9-inch shell	900	11	60	79
packaged mix, double crust	1 (8- or 9-inch shell)	1,485	20	93	141
Pies:					
apple, double crust	1/6 of 9-inch pie	405	3	18	60
custard	1/6 of 9-inch pie	330	9	17	36
lemon meringue	1/6 of 9-inch pie	355	5	14	53
pecan	1/6 of 9-inch pie	575	7	32	71
pumpkin	1/6 of 9-inch pie	370	6	17	37
Pineapple:					
canned, crushed	1 cup	200	1	0	52
juice, unsweetened	1 cup	140	1	0	34
sliced, canned	1 slice	45	0	0	12
fresh, diced	1 cup	75	1	1	19

FOOD	AMOUNT	CAL-ORIES	PRO-TEIN	FAT	CARBO-HYDRATE
Pizza, cheese	1/8 of 15-inch pie	290	15	9	39
Plums, fresh	1 (2 1/8-inch diameter)	35	1	0	9
Popcorn, popped, with oil	1 cup	55	1	3	6
Pork:					
ham, roasted	3 ounces	160	20	8	0
loin chop, uncooked	4 ounces (lean meet only)	210	22	13	0
Potatoes:					
baked	1 medium	220	5	0	51
boiled	1 medium	120	3	0	27
French-fried in oil	10 pieces	160	2	8	20
mashed	1 cup, milk and margarine	225	4	9	35
sweet, baked	1 medium	115	2	0	28
sweet, candied	1 piece (2 1/2 inch x 2 inch)	145	1	3	29
Potato Chips	10 medium	105	1	7	10
Pretzels	10 sticks (2 1/4-inch long)	10	0	0	2
Prune Juice, canned	1 cup	180	2	0	45
Prunes, dried	5 large	115	1	0	31
Pudding, chocolate	5 ounce can	205	3	11	30
Pumpkin, canned	1 cup	85	3	1	20
Pumpkin Seeds dry, hulled	1 ounce	155	7	13	5
Radishes, raw	4	5	0	0	1
Raisins, seedless	1 cup	435	5	1	115
Raspberries, red fresh	1 cup	60	1	1	14
Rhubarb, cooked with sugar	1 cup	280	1	0	75
Rice:					
brown, cooked	1 cup	330	5	1	50
white, enr., uncooked	1 cup	670	12	1	149
cooked	1 cup	225	4	0	50
Rolls, dinner	1 (2 1/2 inch diameter)	85	2	2	14
Salad Dressings:					
blue or Roquefort	1 tablespoon	75	1	8	1
French	1 tablespoon	85	0	9	1
Italian	1 tablespoon	80	0	9	1
Thousand Island	1 tablespoon	60	0	9	2
Salmon:					
baked (red)	3 ounces	140	21	5	0
canned (pink)	3 ounces	120	17	5	0
Sauces:					
barbecue	1/4 cup	80	1	0	16
Hollandaise (w/water)	1 cup	240	5	20	14
spaghetti, canned, tomato puree, canned	l cup	105	4	0	25
white (w/milk)	l cup	240	10	13	21
Sauerkraut, canned	l cup	45	2	0	10
Sausage, pork link cooked	1 link (16 per pound)	50	3	4	Tr
brown and serve	1 link	50	2	5	1
Scallops, breaded, frozen	6 scallops	195	15	10	10
Sherbet	1 cup (2% fat)	270	2	4	59
Shortening, solid	1 tablespoon	115	0	13	0
	1 cup	1,810	0	205	0
Shrimp, French fried	3 ounces	200	16	10	11
canned	3 ounces	100	21	1	1

FOOD	AMOUNT	CAL-ORIES	PRO-TEIN	FAT	CARBO-HYDRATE
Soups (canned soups are prepared according to label directions):					
beef broth	1 cup	15	3	1	0
chicken, cream of (with milk)	1 cup	190	7	11	15
chicken-noodle	1 cup	75	4	2	9
chicken with rice	1 cup	60	4	2	7
clam chowder, New England	1 cup	165	9	7	17
mushroom, cream of	1 cup	205	6	14	15
onion (from packet)	1 cup	20	1	0	4
pea, green	1 cup	165	9	3	27
tomato, clear	1 cup	85	2	2	17
cream of	1 cup	160	6	6	22
vegetable	1 cup	70	2	2	12
Spaghetti, cooked	1 cup	190	7	1	39
w/meatballs & sauce	1 cup	330	19	12	39
w/sauce & cheese	1 cup	260	9	9	37
Spinach, fresh, cooked	1 cup	40	5	0	7
Squash:					
summer, sliced	1 cup	35	2	1	8
winter, baked	1 cup	80	2	1	18
Strawberries, fresh	1 cup	45	1	1	10
Sugar:					
brown	1 cup	820	0	0	212
granulated	1 tablespoon	45	0	0	12
	1 cup	770	0	0	199
powdered (confectioners')	1 cup	385	0	0	100
Sunflower Seeds, dry, hulled	1 ounce	160	6	14	5
Sweet Potatoes	See Potatoes				
Syrup:					
maple	2 tablespoons	122	0	0	32
table (corn)	2 tablespoons	122	0	0	32
Tangerine	1 (2 3/8-inch diameter)	35	1	0	9
without sugar or cream	1 cup	0	0	0	0
Tomato Juice, canned	1 cup	40	2	0	10
Tomatoes:					
canned	1 cup	50	2	1	10
fresh	1 medium	25	1	0	5
Trout, broiled	3 ounces	175	21	9	0
Tuna:					
canned in oil	3 ounces	165	24	7	0
canned water pack	3 ounces	135	30	1	0
Turkey, roasted (white and dark meat)	3 ounces	130	18	5	3
Turnips, white, cooked	1 cup	30	1	0	5
Veal:					
cutlet, broiled	3 ounces	185	23	9	0
Vegetable Juice, canned	1 cup	45	2	0	11
Vinegar	1 tablespoon	0	0	0	1
Waffle	1 7-inch waffle	205	7	8	27
Walnuts, chopped					
black	1 cup	760	30	71	15
English	1 cup	770	17	74	22
Watermelon	1 wedge, 4 inch x 8 inch	155	3	2	35
Yeast:					
Baker's, dry, active	1 package	20	3	0	3
Brewer's dry	1 tablespoon	25	3	0	3
Yogurt	See Milk				

Glossary of Terms

The following terms can be found on the listed page in the Student Manual, unless otherwise noted. ◆ indicates pages which you may receive from your teacher.

active transport	◆ 135
adenosine triphosphate (ATP)	63
amino	◆ 61
amino acids	◆ 61
anabolism	52
base	◆ 67
biomolecules	24
carbohydrates	33
carboxyl	◆ 61
carrier protein	◆ 135
catabolism	52
catalyze	◆ 81
chemical reaction	20
chlorophyll	17
closed system	17
daughter cell	103
deoxyribose	◆ 67
disaccharide	24
electron	41
elements	41
enzymes	◆ 81
eukaryote	92
evolution	82
facilitated diffusion	◆ 133
fats	◆ 65
food	22
glycogen	33
glycolysis	63
growth	99
indicators	28
intermediate	53
lipids	33
liposome	84
metabolism	53
mitosis	103
molecule	41
monosaccharides	◆ 63
neutron	41
nucleic acids	◆ 67
nucleotides	◆ 67
nucleus	41
nutrients	28
organic	52
oxidized	53
passive transport	◆ 133
peptide bond	◆ 61
phloem	17
phosphate group	◆ 67
phospholipids	◆ 66
photosynthesis	16
polysaccharide	24
positive control	19
prokaryote	92
protein	◆ 61
proton	41
pyruvate	63
respiration	65
ribose	◆ 67
saturated	◆ 65
simple diffusion	◆ 133
solar energy	17
starch	33
steroids	◆ 65
stomata	24
unsaturated	◆ 65
variable	19
xylem	17